Excel HSC

FOOD TECHNOLOGY

Get the Results You Want!

DONNA KERR
&
SANDRA WILSON

Reprinted with HSC cards 2004
Reprinted 2005, 2008
Revised edition for HSC syllabus changes 2010
Reprinted 2012, 2015, 2017
Updated 2019
Reprinted 2022

ISBN 978 1 74125 316 0

Pascal Press
PO Box 250
Glebe NSW 2037
(02) 8585 4050
www.pascalpress.com.au

Publisher: Vivienne Joannou
Project editor: Mark Dixon
Edited by Catherine Etteridge and Leanne Poll
Typeset by Grizzly Graphics (Leanne Richters)
Cover by DiZign Pty Ltd
Photos by PhotoDisc, stockByte, Corel and iStock.com
Printed by Vivar Printing/Green Giant Press

Students

All care has been taken in the preparation of this study guide, but please check with your teacher or the NSW Education Standards Authority about the exact requirements of the course you are studying as these can change from year to year.

Contents

Introduction and Course Outline

Introduction

Please spend 15 to 20 minutes reading the following information. (So often readers skip this part of a book!) There are no magical answers as to how each individual should study. Study is more effort-dependent than time-dependent. However, with so much to do, it makes sense to study in a way that makes effective use of your time. Anyone who has studied could list time wasters—the things that you do to put off the actual hard slog of revising notes and trying to make sense of the subject.

You are more likely to be successful if you:

- Understand the structure and ensure you meet all requirements of the 2 Unit Food Technology course.
- Develop and practise good study and work habits to achieve the outcomes of the Food Technology course. Apply practical examples of coursework to your daily lives. It will reinforce what you are learning. This is much easier when you enjoy the subject!
- Develop techniques to ensure you can demonstrate your knowledge and skills in exams and assessment tasks.
- Complete the activities as they are designed to ensure you can do the 'learn to' section of the syllabus.

The structure and requirements of the 2 Unit Food Technology course

The Preliminary course—syllabus overview

The Preliminary course consists of core content that should be covered in 120 indicative hours.

The Preliminary course incorporates the study of:

- Food Availability and Selection (36 indicative hours)
- Food Quality (48 indicative hours)
- Nutrition (36 indicative hours).

The Preliminary course contains content that is considered assumed knowledge for the HSC course.

The HSC—syllabus overview

The new HSC is about developing independent lifelong learners. You will be rewarded for showing initiative to complete this course to the highest standard possible.

The HSC course builds upon the Preliminary course.

The HSC course incorporates the study of four core strands. The core topics constitute 120 indicative hours and include:

- The Australian Food Industry (30 indicative hours)
- Food Manufacture (30 indicative hours)
- Food Product Development (30 indicative hours)
- Contemporary Nutrition Issues (30 indicative hours).

Note: During the course students will learn to:

- Collect, analyse and organise information
- Communicate ideas and information
- Plan and organise activities
- Work individually and in teams
- Solve problems
- Use mathematical ideas and techniques
- Use technology by participating in practical activities.

HSC assessment

Your HSC Food Technology course result is based on:

- An (internal) assessment mark submitted by your school, and
- An (external) examination mark derived from the HSC Food Technology Examination.

Assessment will be based on the course objectives and outcomes identified in the NESA syllabus document.

Your HSC results

A course report containing a performance scale with bands describing standards of achievement will be given. Your indicated position on these bands will determine your result.

Examination results are no longer obtained by students competing against each other to receive a ranked order. Students are now being measured against standards that reflect achievement of course outcomes. Students who perform to the same standard should get the same result, irrespective of which year they sit the HSC Food Technology Examination. Close analysis of the achievements of students in the examination and internal assessment will be used to construct the descriptions of student achievement that make up each performance level. Students should be familiar with the performance band descriptions. The standard of work required increases or improves as the band number increases.

External examination

The examination will consist of one paper of three hours' duration plus 5 minutes reading time. It will have FOUR sections.

There will be approximately equal weighting of each of the four core strands across the examination as a whole. Questions may require students to integrate knowledge, understanding and skills developed through studying the entire course, rather than focusing on a particular core strand.

Section I (20 marks)

- There will be 20 multiple-choice questions.
- All questions will be compulsory.
- All questions will be of equal value.
- Questions are answered on a separate answer sheet.
- Questions will be based on the HSC core strands.

Section II (50 marks)

- There will be approximately 6 short-answer questions.
- Questions may contain parts.
- There will be approximately 14 items in total.
- At least four items will be worth from 4 to 6 marks.

Section III (15 marks)

- There will be one structured extended-response question.
- The question will have two or three parts, with one part worth at least 8 marks.
- The question will have an expected length of response of around four pages of an examination writing booklet (approximately 600 words) in total.

Section IV (15 marks)

- There will be one extended-response question.
- The question will have an expected length of response of around four pages of an examination writing booklet (approximately 600 words) in total.

Source: *Assessment and Reporting in Food Technology Stage 6*, p. 8, © 2017 NSW Education Standards Authority

Preparation for the exam involves trying to lengthen your concentration span. It is important not to waste time or daydream during the exam.

Students should note:

In section IV the examination rubric states:

In your answer you will be assessed on how well you:

- Demonstrate knowledge and understanding relevant to the question
- Apply course concepts to food technology issues
- Communicate ideas and information using appropriate terminology and relevant examples
- Present a logical and cohesive response.

Source: *HSC Food Technology Examination © 2018 NSW Education Standards Authority*

Internal assessment

Components, weightings and suggested tasks for the HSC course internal assessment are shown below. These are specified in the syllabus.

Component	**Weighting**
Australian Food Industry	25%
Food Manufacture	25%
Food Product Development	25%
Contemporary Nutrition Issues	25%
Total Marks	**100%**

Internal assessment could include food preparation and presentation exercises, experiments, research assignments, debates, oral presentations, case studies and industry reports.

Assessment tasks should be set to ensure that:

- 20% of marks are based on knowledge and understanding
- 30% of assessment marks should be earned by research analysis and communication
- 30% of assessed marks are based on experimentation and preparation
- 20% of marks are calculated by activities which require students to design, implement and evaluate.

Slight modifications of the syllabus may occur from time-to-time. Everyone can access the latest version of the syllabus through the NESA website:
www.educationstandards.nsw.edu.au

It is also worth accessing examination reports from previous exams from this site. These detail many of the common errors that students make.

Summary of external and internal HSC assessment

External examination	Mark
Section I Objective-response questions	20
Section II Short-answer questions	50
Section III Candidates answer one structured extended-response question	15
Section IV Candidates answer one extended-response question	15
	100

Internal assessment	Weighting
Knowledge and understanding of course content	40
Knowledge and skills in designing, researching, analysing and evaluating	30
Skills in experimenting with and preparing food by applying theoretical concepts	30
	100

Source: *Assessment and Reporting in Food Technology Stage 6*, pp. 7–8, © 2018 NSW Education Standards Authority

Skills development

It is important that you are familiar with **ALL** the skills listed in the course. The activities you undertake should lead to and enhance your understanding of the main ideas in the course. You are expected to have employed all the skills at least once. You may be asked in an exam, for example, to explain how you have carried out one of the activities in the 'Students learn to' section of the syllabus.

Using the internet—one of the skill areas that you may need to develop is your ability to use the internet efficiently and effectively. While this guide gives you some addresses, it is essential that you develop expertise in using search engines. The use of + and – signs to help clarify search terms is often useful. Making the most of either searching by categories or browsing can also aid in information gathering.

Resources—make use of a range of resources and your teacher to ensure you are focused on essential course content. This study guide has been written so that it closely aligns with the syllabus.

The new Students Online website (https://studentsonline.nesa.nsw.edu.au)provides excellent detailed advice about study and exam techniques, as well as useful links to information sites for most subjects.

Hints for developing good study and work techniques and habits

Below are some hints for developing good study techniques:

1. Establish an appropriate **place** to study—somewhere quiet, free from interruption, comfortable (but not too comfortable) and organised.

2. Practise **self-discipline** and **self-motivation.** This can be linked to establishing clear long-term and short-term goals. Establish a goal, for example, 'Plan a strategy for the marketing of a specific food product'. It should be clear when you have achieved this goal so you can give yourself a small reward as well as having the satisfaction of achieving your goal.

   Your motivation to study will improve when you link studying to the confidence that you can achieve goals. If you are struggling with deciding a career path, try to focus on the long-term goal of success in the HSC and/or obtaining a good UAI. Even if you don't want to follow a career in food technology, doing well in this course will produce benefits that relate to many career paths and your life in general. You will develop skills in lifelong learning and become an informed citizen on important issues in today's society.

3. Prepare a **study timetable**—do not study one topic for too long, and try to revisit topics with a reasonable degree of frequency. You have to balance the demands of school assessment tasks with revision and preparation for the external examination. Your timetable should include preparation for the trial and other assessment tasks.

4. Establish a **regular time** for study.

5. **Take care of yourself**—set aside time for mental relaxation, keep physically fit, eat regular meals, get plenty of sleep. Your state of mind is also important. If you are stressed, or don't believe you can learn, or can't see the point of what you are learning, you won't learn well.

6. Reflect on how you learn best and what is working for you. **Remember:** we are all unique in the way we prefer to take in information and learn. There is no one right way. If you know how you best learn, you can make your study time more productive.

7. Show yourself or someone else that you understand what you've learned. Just reading through notes is not always the best way. Doing something with the information helps it stick in your memory. Developing your own bank of questions and answers

works well for some people. To make sense of information you need to:

- **Read it**—see it
- **Hear it**—say it aloud
- **Write it**—do something with the information.

8. Memorise one or two key facts so the rest of what you've learned comes flooding back. Put the key points down in your own words. Put the key ideas into tables or draw diagrams. Some people use associations such as acronyms to help remember. You might write key words on cards or post-it notes, or draw a learning map of all the key concepts. The important thing is that effective learning occurs when you combine **seeing**, **hearing**, **and doing**.

Use the information in this book as a guide to organising your own notes and preparing your own summary. Use the answers/explanations to modify your notes or resummarise your summarised notes.

How to use this study and revision guide

The purpose of this book is to:

1. Provide you with a course outline that will help you study and make sense of the subject. The outline includes a variety of features including key points, summary tables, written notes, diagrams and key terms.
2. Give suggestions of the ways you can approach many of the mandatory practical and information skills activities in which you will need to be able to demonstrate competence in the course.
3. Provide you with practice questions and a sample trial exam. (The more you can do, the better, and try doing some of the questions under exam conditions.)

Each chapter has:

1. **A list of outcomes**.
2. **An extended outline** covering all the important concepts and facts from the course syllabus. These are covered in the order in which they appear in the syllabus. To help you with quick pre-exam revision, the main ideas for each topic are highlighted in boxes, tables, or diagrams, or printed in bold print or capital letters.
3. At the end of each chapter there is a **checklist** of key terms for revision, and a set of **practice questions**—multiple-choice questions with answers and explanations, and short-answer and longer-response questions with answers and explanations. The answers to Activity 1 in each chapter can be used as a glossary for memorising and revising key terms and concepts.

Also included are:

- A **sample HSC Examination** in Chapter 5 with answers and comprehensive explanations
- A **glossary of key terms** for the HSC core modules on page 138.

Note: The HSC Food Technology course introduces you to many new words. It is a fairly 'wordy' course. Words that you do not understand, may be technical words (for example, prototype, malnutrition, 'cradle to grave') or everyday words that also have a special scientific meaning (for example, environment, preservation, functional). You need to list these words and satisfy yourself that you understand what they mean before going too far with the rest of your revision.

Hints for answering exam questions

1. **Read all parts of each question before you begin**. Work out how many parts need to be answered. Also check the mark value of each question. Some questions may be worth 15 marks but consist of a number of

parts. At first these structured questions may seem difficult, but it is essential that you attempt them, even if you do not feel confident. Not attempting questions is just throwing away marks.

2. **Answer all parts and answer in the correct order and in the correct space**. Underline words in the question that tell you what is being asked.

3. **Be familiar with the language of exams and the language of the syllabus**.

(a) Questions containing the following instructions require short answers of only a few words or one or two sentences:

- Give one reason, define, describe, identify, outline, recall, recount, summarise, list, name, state.
- Learn the meaning of these words from the NESA definitions.
- List THREE factors; name TWO functions; what are FOUR features; give TWO examples and so on.

(b) Answers to questions containing the following terms usually require slightly longer answers:

- Assess, critically analyse, justify, classify, compare, contrast, deduce, demonstrate, discuss, evaluate, examine, explain, extract, extrapolate, interpret, investigate, predict.
- Understand these instructions and respond appropriately.

At times you may have to bring together ideas from throughout a module for your answer. Remember: your answer should be **comprehensive but concise**.

4. **Decide how best to organise your answer.**

- Full sentences may not be necessary and are often a waste of space. This is not like an English essay.
- Organise your thoughts logically—use flowcharts and accurately labelled diagrams where appropriate.
- Respond in relation to the space provided.
- Do not rewrite the question.

5. **Answer only the question asked.** This involves taking the time to read the question carefully and determining exactly what it means. HSC Examinations are set on a criterion or standards-referenced approach. This means that there will be a full range in the degree of difficulty of questions. Generally within sections or multiple-part questions, the easier questions will come first.

6. **Do not expect the examiner** (the person(s) marking your answer) **to read your mind**. Don't say to yourself 'They will know what I mean', and don't expect examiners to fill in the gaps in your answer. State all relevant facts and give complete explanations where appropriate.

7. **Answer the question as a person who has studied the subject** (not from general knowledge). Don't use slang. Remember: the question is testing specific knowledge of the course; if you think you can answer it using general knowledge, you are probably missing the point of the question.

8. **Plan your timing and approach to the exam ahead of time**. You have 180 minutes to achieve a maximum of 100 marks. That means approximately 30 minutes to complete the multiple choice section, 90 minutes to complete the structured questions, 30 minutes for the Section III question and 30 minutes for the Section IV question.

9. **Do not panic if you think you are running out of time**—simply jot down the essential points on the answer booklet and come back later if time permits.

10. **Read over your answers to all sections after you have finished**.

Food Technology HSC Course — Stage 6

Core Strands (100% total)

The Australian Food Industry (25%)
- Sectors of the Australian food industry
- Aspects of the Australian food industry
- Policy and legislation

Food Manufacture (25%)
- Production and processing of food
- Preservation
- Packaging, storage and distribution

Food Product Development (25%)
- Factors which impact on food product development
- Reasons for and types of food product developments
- Steps in food product developments
- Marketing plans

Contemporary Nutrition Issues (25%)
- Diet and health in Australia
- Influences on nutritional status

1–Australian Food Industry

The Australian food industry has developed in response to changes in our physical, social, technological, economic and political environment. This is evident in the structure, operations and products of the Australian food industry. The industry contributes significantly to the gross domestic product and is a major employer.

Outcomes

On completing this chapter a student should be able to:

1. Examine the nature and extent of the Australian food industry (H1.2)
2. Evaluate the impact of the operation of an organisation within the Australian food industry on the individual, society and environment (H1.4)
3. Investigate operations of one organisation within the Australian food industry (H3.1).

Activity 1 **A PAGES 28–29** (**Note: A PAGES 28–29 refers to the answer pages for this activity**.)

Following is a list of terms you should know as a result of completing this module.

Term	Meaning
Agriculture	
Agri-food chain	
ANZFA	
AQIS	
Aquaculture	
Biotechnology	
Catering	
'Clean green'	
Embargo	
Fisheries	
Genetic engineering	
Legislation	
Manufactured foods	
Multinational	
Policies	
Primary industries	
Primary produce	
Processing	
Quota	
Recall	
Subsidy	
Tariff	
Technology	
Value-added	

The HSC Examination

In the HSC Examination students should expect to answer questions on the Australian Food Industry to the value of approximately 25% of the paper. Questions may require students to integrate knowledge, understanding and skills developed through studying the entire course. It is therefore difficult to predict the weighting of marks in each section of the paper. The paper may include:

- Multiple-choice questions
- Short-answer questions (which may include parts)
- A structured extended-response question, which includes two or three parts with one part worth at least 8 marks

OR

- An extended-response question (approximately 600 words).

1.1 Sectors of the Australian food industry

Students must be able to **identify** the sectors within the Australian food industry and should have completed at least one case study whereby they have researched **recent developments** in one sector of their choice from the following:

- Agriculture and fisheries
- Food processing and manufacturing
- Food retail
- Food service and catering.

Agriculture and fisheries

This sector includes **agriculture**, which is the cultivation of land to produce crops and/or animals, as well as **fisheries**, which involves the cultivation of various cold blooded, aquatic species usually for a commercial or scientific purpose. Key features of the agriculture and fisheries sectors in Australia include:

- Production of plants and animals.
- Historically, Australians have lived 'off the farmer's back' because food production has been important to the Australian economy.
- Availability of fertile land and unpolluted waterways (salt and fresh).
- Climate harsh but **technological development** has been able to compensate for extreme conditions (e.g. irrigation, genetic engineering, farming equipment).
- **Primary industries** contribute significantly to the Australian economy, providing a safe, reliable food supply and a significant **export** market.
- **'Clean green' image** of Australian produce ensures it is highly regarded internationally.
- **Crops** include cereals, especially wheat and more recently rice; also vegetables, fruits, and nuts.
- **Animal production** includes lamb, beef, pork, poultry; also small-scale production of venison, kangaroo and emu for the gourmet food industry.
- **Animal products** include meats, milks, cheeses, yoghurts, cream, butter and cooking fats (of animal origin, such as butter and lard).
- **Fishery products** include fish (salt and fresh water), prawns, yabbies, oysters, lobster, scallops, mussels, etc. Many species are caught in 'the wild' (i.e. rivers, oceans, lakes), however, increasingly **aquaculture** or controlled breeding programs (e.g. fish farms) are providing reliable supplies at stable prices.

Food processing and manufacture

Australia is now heavily involved in the **value-added** area (processing of goods to increase their selling price) of food production. The move towards this stage of food production has enhanced the economic value of the food industry for Australia. Key features of food **processing** and manufacture in Australia include:

- Traditionally, Australia focused on **primary produce** but increasingly the economy has benefited from processing and manufacturing 'our' produce.
- **Value-added** foods (e.g. wheat manufactured into breakfast cereals, flour or bread) increase opportunities to provide employment for Australians, and the increased revenue (from the higher priced foods) remains in Australia.
- Processing and manufacture in Australia must compete with cheap **labour markets** overseas.
- The cost of labour in Australia has meant that food processing and manufacture has had to become **mechanised** to ensure processing is economically viable.
- Many of the food manufacturing industries are **multinational** organisations.
- The industry is **diverse**, from household production to large multinational organisations.

Food retail

Retail involves the selling of food at all levels. The manner in which food is sold varies enormously depending on the foods sold and the **distribution channels**. There is increasing interest in purchasing Australian food within Australia and overseas. In food retail, the following trends have emerged:

- Customers are using the 'corner' convenience store less, in favour of supermarket **one-stop shopping**.
- **Specialty stores** are a feature of Australian shopping (e.g. bakers, butchers and delicatessens).
- Food sales occur in a variety of **settings** such as theatres, transport terminals, dispensing machines and service stations.
- **Lifestyle** changes, such as changing work patterns, leave less time to 'shop around'.
- Greater motor vehicle usage has increased mobility and opportunity to transport **bulk purchases**.
- More convenience foods are being purchased.
- Storage facilities are improving.
- More recently, the availability of **internet shopping** has provided an efficient alternative to in-store shopping.

- Self-serve shopping increases **impulse buying**.
- Growing popularity of ready-to-eat meals and **value-added** products save time and energy (suit busy lifestyles). Often referred to as '**home meal replacements**', they provide a fast alternative for the busy consumer, as the food can be ready to eat or require minimal preparation. Convenience is the key feature, with many high quality nutritious choices available.
- **Health-conscious** consumers have renewed interest in 'fresh food'.
- Rapidly increasing overseas market (especially Asia) has resulted from the '**clean green**' image of Australian foods and from the value of the **Australian dollar** in the international marketplace.

Food service and catering

The food service and **catering** sectors encompass all aspects of commercial food production and service (hospitality industry). In this sector the following trends have emerged:

- Australians are eating more meals away from home than at any time previously (at least one meal in four).
- Some foods are available through institutions, such as hospitals, prisons, armed services and boarding schools. These are often referred to as **non-commercial meals** because they are a feature of another major service (the reason for the institution—health care, security, etc.)
- Other organisations exist for the sole purpose of providing meals as a commercial enterprise, such as restaurants, reception centres, clubs, snack bars and take-away outlets, and making a profit from the sales of their food items.
- Caterers prepare, cook and serve food. This service is increasingly utilised in today's society as people share food (which they have not prepared) in a variety of social situations.
- **Multinational** retail chains compete aggressively for the 'fast-food dollar' (e.g. McDonald's, Pizza Hut, KFC).
- Home delivery service has become an important feature of the fast-food industry.

Activity 2 A PAGE 29

Complete the following table:

Development	Sector/s	Explanation
One-stop shopping		
Irrigation		
Convenience meals		
Aquaculture		
Home delivery service		
Genetic engineering		
Value-added foods rather than primary produce		
internet shopping		
Breeding of lean meats		
Product scanning		
Pollution control		
Australia's 'clean green' image		
Production of low-fat animal products		

Activity 3 A PAGE 30

Complete the following summary table indicating foods/meals that reflect each sector:

Sector	Foods/meals that reflect that sector
Agriculture and fisheries	
Food processing and manufacture	
Food retail	
Food service and catering	

Emerging technologies in food production, manufacturing and packaging

New technologies in the areas of food production, manufacturing and packaging are constantly evolving in the Australian food industry.

Genetic engineering

The process of genetic engineering involves biologists removing a gene from a living organism and inserting it in another organism; for example, inserting a **disease-resistant gene** into a plant. This work has increased understanding about how plants grow and how plant genes work.

This level of **biotechnology** allows characteristics of foods to be improved by transplanting genes from one living thing to another. However, the genetic engineering of food is an ethical and controversial issue. Concern has been expressed about interfering with nature and animal rights.

Gene surgeons have developed different strains of yeast for brewing and baking. Genetically engineered potatoes (resistant to infection by the leaf-roll virus that discolours the potatoes and reduces yields by up to 30%) are also being used in the snack food industry. Cereals have been engineered to allow them to be sprayed for weeds without harming the cereal plant. Genetically engineered grapes used in winemaking are able to be picked without discolouration and drying.

Genomics is the science of gene mapping. The technology has developed rapidly in recent years and has made it possible to make complete gene maps of organisms. It is expected to be of great value for breeding. In conventional breeding the search for best crossing partners is a tedious and time consuming trial-and-error procedure. With genomics it is possible to select breeding partners on the basis of gene maps. According to experts on genomics, it can reduce the number of crossing trials very substantially.

Transgenic meats are also being developed. For example, in pork production Enviropigs are able to digest the plant phosphorous more efficiently, which means there is less phosphorous—up to 60% less than ordinary pigs—in their waste. That, in turn, means less phosphorous will leach from pig manure, a major fertiliser source for farmers, into freshwater lakes and streams where it can trigger vast algal blooms and kill fish. Alternatively, growth hormones have been inserted into embryos to produce pigs with really lean flesh, which reach their market rate faster.

Packaging

Technological advances have meant packaging is becoming more environmentally friendly and cost efficient. Advance in packaging design and materials have resulted in products that are more efficient and have also contributed to more efficient production methods (see Chapter 2, page 56 for more information on packaging).

Ecologically sustainable production methods

Ecologically sustainable food production methods have become increasingly important in the marketplace and a key feature of food production and manufacturing.

Organic farming

Sustainable agriculture means, by definition, agriculture that does not deplete natural resources and does not use harmful, artificial substances that cumulate in the environment. It is the only kind of agriculture that is feasible in the long run. A common word for it is 'organic farming'.

Organic farming aims to maintain or improve the fertility and the level of organic matter in the soil. Generally yields are lower than they are in conventional farming because all production is chemical-free, making pest control less effective. Farmers are required to have their soil and farming practices evaluated before selling their products as organically grown. Many consumers believe organically grown foods are superior in quality and are willing to pay more for them.

Also as a result of **lower productivity**, organic foods tend to be more **expensive**. Consumers are increasingly concerned about the environment and are anxious to consume foods that are free of chemicals. Farmers are required to have their soil and farming **practices** evaluated before selling their products as organically grown. The **Australian Quarantine Inspection Service (AQIS)** is responsible for ensuring the food meets the required **food standards** of organic farming prior to export. Despite the need to meet strict standards, the number of organic farmers is increasing as a part of the Australian food industry sector, agriculture and fisheries.

Activity 4 A PAGES 30–31

(a) **Investigate** and **describe** an emerging technology in ONE sector of the Australian food industry.

(b) Complete the following table discussing the potential risks and benefits of using emerging technologies in food production and manufacture. The first entry has been done as a guide.

Emerging technology	Risks	Benefits
Genetic engineering	■ **Imprecise technology**—A gene can be cut precisely from the DNA of an organism, but the insertion into the DNA of the target organism is basically random. As a consequence, there is a risk that it may disrupt the functioning of other genes essential to the life of that organism. ■ **Side effects**—Scientists do not yet understand living systems completely enough to perform DNA surgery without creating mutations which could be harmful to the environment and our health. ■ **Widespread crop failure**—Genetic engineers intend to profit by patenting genetically engineered seeds. This means that when a farmer plants genetically engineered seeds, all the seeds have an identical genetic structure. As a result, if a fungus, virus or a pest develops which can attack this particular crop, there could be widespread crop failure. ■ **Threatens our entire food supply**—Insects, birds and wind can carry genetically altered seeds into neighbouring fields and beyond. Pollen from transgenic plants can cross-pollinate with genetically natural crops and wild relatives. ■ **No long-term safety testing**—Genetic engineering uses material from organisms that have never been part of the human food supply to change the fundamental nature of the food we eat. Without long-term testing no one knows if these foods are safe. ■ **Toxins**—Genetic engineering can cause unexpected mutations in an organism, which can create new and higher levels of toxins in foods. ■ **Allergic reactions**—Genetic engineering can also produce unforeseen and unknown allergens in foods. ■ **Decreased nutritional value**—Transgenic foods may mislead consumers with counterfeit freshness. A luscious-looking, bright red genetically engineered tomato could be several weeks old and of little nutritional worth. ■ **Increased use of herbicides**—Scientists estimate that plants genetically engineered to be herbicide-resistant will greatly increase the amount of herbicide use. Farmers, knowing that their crops can tolerate the herbicides, will use them more liberally. ■ **Ecology may be damaged**—The influence of a genetically engineered organism on the food chain may damage the local ecology. The new organism may compete successfully with wild relatives, causing unforeseen changes in the environment.	■ Genetic engineering reduces the costs of production, meaning that the poor can afford more food, and be more self-sufficient ■ Better resistance to weeds, pests and disease ■ Better texture, flavour and nutritional value ■ Longer shelf life, easier shipment ■ Better yield, more efficient use of land ■ Less herbicides and other chemicals ■ Cheaper and safer source of human medicine ■ Pest and disease resistant ■ Herbicide resistant ■ Easier to harvest plants ■ Increased productivity ■ Fewer agricultural and horticultural chemicals needed ■ Fewer chemical residues in food chain ■ Remove undesirable characteristics ■ Design what people want or need ■ Improve or add desirable characteristics ■ Reduce management costs

Emerging technology	Risks	Benefits
Genetic engineering (continued)	■ **Gene pollution cannot be cleaned up**—Once genetically engineered organisms, bacteria and viruses are released into the environment, it is impossible to contain or recall them. Unlike chemical or nuclear contamination, negative effects are irreversible.	
Genomics		
Transgenic meats		
Packaging technology		
Organic farming		
Additional technology investigated		

1.2 Aspects of the Australian food industry

Level of operation

In the Australian food industry, the level of **technology** (mechanisation and often computerisation) used by an operation generally increases with the size of the operation. More people are employed. Larger companies tend to have increased political influence over a range of food issues. In Australia, many successful large business are being taken over by **multinational** companies.

Level	Description	Features
Household	Home produce or manufactured products	Always small-scale with low levels of technology
Small business	Often family business or partnership, often less than 20 staff members	Tends to be local (e.g. local Chinese restaurant)
Large business	One which operates across the State or even nation (e.g. United Dairies)	Often a number of owners servicing a relatively broad community
Multinational	A food company which operates in several countries (e.g. McDonalds)	Large-scale and global

Research and development

Research and development is the process used to create new products. Competition drives food manufacturers to create new products for the purpose of winning market share. In the Australian food industry, features of research and development include:

- The marketplace is expanding and consumer groups are very diverse.
- There is continual pressure on the food industry to provide **innovative** products to meet the changing needs of consumers.
- Government groups like the Commonwealth Scientific Research and Industrial Organisation (**CSIRO**) undertake significant food related research.
- Many food businesses undertake research in attempts to develop new food products that will help them to gain and maintain the market edge. For example, **high-yield produce**, nutritionally improved foods, extended shelf-life products and environmentally friendly products.

Quality assurance

Quality assurance is the responsibility of every person involved with the production of a food product (raw material→processed product→retailer). Features of quality assurance include:

- Food quality is **guaranteed** through strict quality control procedures in order to increase consumer confidence.
- In Australia, consumers have the support of **consumer organisations** (Consumer Affairs Bureau, for example) to ensure product satisfaction.
- Most organisations in the food industry have introduced a **customer complaint procedure**, with many offering a 'money back' guarantee.
- In order to guarantee high standards, **quality control processes** are employed.
- **HACCP** (Hazard Analysis Critical Control Points) is an international approach to quality assurance. This process requires:
  - assessment of the hazards
  - identification of critical control points
  - set standards at each control point
  - control point monitoring
  - clear advice about procedural changes to ensure standards are met
  - recording of operations/standards to track variations
  - supervision and checking.

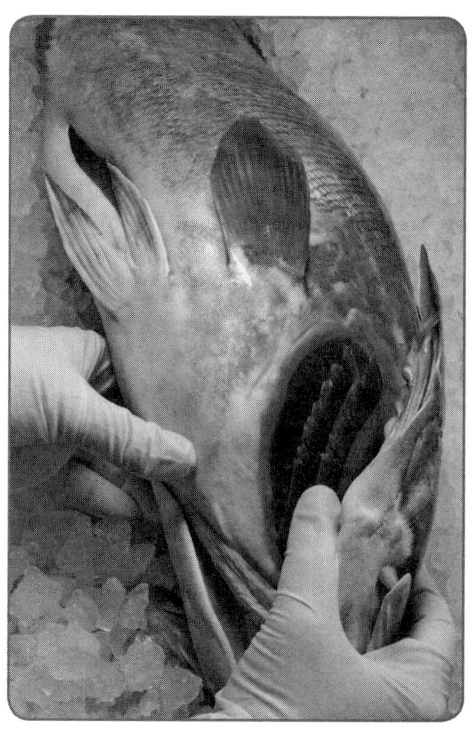

Consumer influences

The changing nature of Australian society has resulted in demands for different foods. Some of the changes in society include:

Social trends	Changing demand	Examples of products
Single parent families increasing	■ Low-cost nutritious foods	■ Cheaper cuts of meat ■ Fruits and vegetables in season
Dual income families increasing	■ Value-added quickly prepared meals	■ Home meal replacements
More mobile population	■ Increased interest in takeaway foods	■ Takeaway meal (e.g. fish and chips, Chinese food)
Busier lifestyles	■ More 'instant' foods	■ Prepared sauces ■ Frozen vegetables ■ Meats which can be prepared quickly by grilling or frying
More time spent away from home	■ Increase in consumption of foods prepared and eaten outside the home	■ Restaurant meals
Smaller family size	■ Less use of cheaper, filling foods	■ Less use of breads ■ Less use of cereals
Improved technology	■ Use of equipment which increases the efficiency of food storage and preparation in the home	■ Microwaves ■ Home freezers ■ Blenders
Higher standard of education	■ Better understanding of diet-related health issues	■ Foods low in fat, salt and sugar ■ Fibre-rich foods ■ Fresh as opposed to processed foods
Increased life expectancy	■ Many elderly people live alone	■ Single serves or small portions are important in packaging and sales
Greater concern for environmental issues	■ Methods of production which are environmentally friendly	■ Organic foods ■ Packaging which is reduced, reusable and/or recyclable

The food industry is responding to the changing needs of consumers with the production of food, which is:

- **Varied** in cost from generic to high quality primary produce to gourmet
- Quick and simple to prepare
- Offered through a wide **variety of food outlets** (food 'on the run') which are easily accessible
- Able to be **heated and served**—this includes meals as well as broad choice of partly prepared foods
- Packaged in **single-serve** portions
- **Appropriate to lifestyle** and available in the food service and catering industry; for example, the breakfast market is becoming very fashionable
- **Microwaveable** or quickly heated
- Prepared by a wide range of **appliances** that offer speed and efficiency in food production in the home (blenders, mixers, bread machines, microwaves, elaborate barbecue facilities)
- **Nutritionally enhanced** (functional foods), such as cholesterol lowering, foods that respond to consumer health issues (low in fat, salt, and sugar, and high in complex carbohydrates).

Value-added convenience foods

As women join the paid workforce in increasing numbers, men are taking a more active role in the shopping for, and preparation of, food. The demand for food that is quickly and easily prepared has increased, as has the percentage of money spent to meet this need. The demand for **increased convenience** has also led to more **flexible shopping hours** and an increase in **self-service**.

Many more meals are eaten **away from home** due to longer working hours, increased flexible working hours and increased shopping hours. Tastes have become more sophisticated and consumers more health conscious resulting in an increase in takeaway foods that are improved in nutritional value and variety.

For meals eaten at home, the trend towards sophisticated 'home meal replacements' provides a useful alternative involving less preparation time and clean up while providing for consumer variety and nutritional awareness.

Home meal replacement

Fresh chilled meal trends have been one of the latest developments in **home meal replacements (HMR)**. These meals are frequently packaged as separate components (e.g. noodles, prepared vegetables and stir-fry sauces) leading to increased innovation in packaging.

Impact on the environment

The Australian food industry has had an effect on the following environmental issues:

- The production of the raw materials (use of **chemicals, salinity, soil degradation**)
- Manufacturing and transport emissions and the use of **fossil fuels**
- Packaging practices
- Waste management
- Transportation.

In response, the following environment-based trends have emerged:

- Greater consideration is given to farming techniques that take into account erosion, soil degradation and the use of chemicals.
- Farming practices have increasingly used science to improve food production without harm to the environment.
- Products that are organically farmed have grown in popularity, strengthening the market for 'clean green' food products and lessening the use of chemicals in food production.
- More careful disposal of manufacturing waste, reduction of packaging and the increased ability to reuse and/or recycle material.

Waste management/packaging practices

Recycling is the **return** of raw materials into the manufacturing process to make other useful products. It is essential because landfill space in urban areas is dramatically decreasing. Examples of how the Australian food industry recycles are listed below:

- 70% of packaging paper is recycled into other products.
- 87% of glass is recovered for refilling or recycling into new packages.
- 65% of aluminium is recycled.
- Mixed plastic waste is converted into timber substitutes such as decking, fencing and outdoor furniture.
- Household waste represents approximately 45% of total waste.
- Material minimisation is a strategy introduced by manufacturers to reduce waste and cost by reducing the weight of packaging but still maintaining packaging strength.
- Some food manufacturers are employing waste management companies to manage waste.
- Waste management involves the cooperation of government, waste management companies and food producers to create waste management that is cost-effective.
- Some packaging can now be made from biodegradable material that will decompose through the action of micro-organisms in a reasonable length of time.

Production techniques

Production techniques involve the following environmental issues:

- **Energy** is used for processing and transportation.
- The main energy used in food manufacture is **non-renewable resources** such as coal and oil.
- Manufacturers need to **reduce energy costs** by investigating more efficient methods of production and transport.
- **Gases released into the atmosphere** during the production of packaging products such as glass, for example, contain more carbon dioxide than is released during the production of plastic packages—and plastic products are also recyclable.

Solutions to some of the associated problems include:

- **Reusing waste materials**; for example, some food manufacturers reuse water from processing for irrigation of plant gardens

- **Recycling** waste materials such as packaging
- Production of **lighter and more efficiently shaped packages** so that more can be transported at the same time.

Pollution comes in many forms: air, land, water and noise. To reduce the impact of pollution, manufacturers have formed waste and pollution **reduction strategies and recycling programs.** Strategies include:

- Reusing and recycling **water**
- Reusing **effluent** and other waste
- Minimising the **use and weight** of packaging
- Converting refrigeration systems from CFC systems
- Creating by-products from waste (e.g. husks and cobs from frozen corn are used for cattle feed).

The Australian packaging industry has **eliminated chlorofluorocarbons** (CFCs) and replaced them with hydrocarbons. Multinational companies such as McDonalds have replaced Styrofoam packaging with **paper** to reduce the emission of harmful environmental gases, and to reduce landfill as paper is recyclable.

Transportation

The type of transportation used in distribution (air, water, rail or road) is dependent on the type of product, the distance to be covered and the nature of the product. Regarding the impact on the environment, food manufacturers are aware of the flow-on in costs. While road and air is quick and efficient, rail is certainly more environmentally friendly. Many food companies are establishing warehouses close to rail lines to make use of this mode of transport. In addition, transport companies have fitted filters to the exhausts of transport vehicles to reduce gas emissions.

Impact on the economy

Generation of profit

The Australian food industry contributes significantly to the Australian economy:

- Food accounts for 46% of all retailing turnover in Australia.
- Many new industries were established in the late 1980s and 1990s to take advantage of emerging market opportunities. Crops in the fruit and vegetables industry, such as Asian vegetables, nashi pears, lychees, olives and herbs, were introduced. New aquaculture activities, such as the farming of Atlantic salmon and growing out of wild-caught southern bluefin tuna, were established.
- Domestic and international markets have recognised the food value of Australia's indigenous flora and fauna. Kangaroo and crocodile, for example, are now accepted meat products. The 'bush foods' industry has worked to integrate a wide range of native products into the Australian food industry.
- The processed food and beverage industry is Australia's largest manufacturing industry with a turnover of more than $71.4 billion in 2005–06.
- Australia's fifty largest food and beverage corporations produce almost three-quarters of the domestic industry's revenue. Supermarkets and grocery outlets continue to capture the majority of food sales in Australia, with around 60% of the value of total food and liquor retailing in 2006–07.

- Meat and grains have consistently been the two largest export categories, with meat accounting for 30% of the value of food exports in 2006–07 and grains nearly 15%. Wine and dairy exports have also grown significantly in recent years, with wine accounting for nearly 13% of exports in 2006–07, and dairy nearly 10%.
- Australia's major markets for exports are Japan and the United States, making up 20% and 13% respectively.
- The food industry is expanding enormously because of the rapid growth in the tourism industry (food service and catering).

Changes in employment

The rapid changes and improvements being made to manufacturing technologies, package design and materials, as well as growth in convenience foods and increased food choice, have provided many opportunities for **employment**:

- The food and beverage sector accounts for more than 17% of employment in the Australian manufacturing sector.
- The processed food and beverage industry makes a significant contribution to rural and regional Australia, with over 40% of food processing employment occurring in non-metropolitan areas.
- The industry makes a significant contribution to the economies of regional areas in Australia through employment, business and service opportunities. There were around 191 400 people employed in food and beverage manufacturing in Australia in 2006–07.
- Small operations tend to be more labour intensive due to the cost of mechanisation.
- Food manufacturers tend to have a flow-on effect in the wider economy by engaging services and resources from around them.
- Employment opportunities are increasing in areas of food product development, materials research, package design, labelling, waste management and food processing.
- Multinational companies employ many staff to produce a variety of product lines. For example, Mars Incorporated has grown into a company of global scope with six business segments (Chocolate, Petcare, Wrigley Gum and Confections, Food, Drinks and Symbioscience) generating annual revenues of $30 billion. In Australia Master-Foods, a subsidiary of Mars, is a major employer. MasterFoods today offers over 500 products in categories as diverse as mustards, marinades, relishes, sauces, herbs and spices, pasta and stir-fry sauces, and Promite spread. It exports to New Zealand and throughout the Asia-Pacific region as well as supplying the food service and industrial sectors.

Impact on society

Food production influences the way society lives. The changing lifestyles of Australians have also impacted on food production and food choices.

- The changing mix of Australian society (migration and travel) has increased food choices, making multicultural foods a large part of Australian eating.
- In the last 60 years there has been an emergence of 'Australian cuisine' flavoured by European and Asian influences; however, the 'fresh produce' trademark of Australia has been dominant.
- The vast variety of foods available in Australia makes choice our biggest problem. The food industry must support consumers in their choices with easily understood labelling, clear nutritional information and informed advertising.

Lifestyle changes

There is an interrelationship between lifestyle changes and developments in the Australian food industry:

- Production of convenience packaging and convenience foods has enabled Australians to enjoy increased **leisure time**.
- Increase in women working, and therefore less time for food preparation, has resulted in improved cooking technologies such as convection microwaves and the development of home meal replacements.
- Food solutions are being produced to suit family members that may eat at different times or who have a wide variety of tastes. Examples include single-serve and fully prepared microwave meals.
- Increased awareness of health has seen the production of **health foods** such as low-fat, high-fibre foods.
- Some manufacturers are attempting to cater to the **busy-lifestyle person** by offering products such as breakfast bars, breakfast shakes and stir-fry kits.
- There is an increased variety of food choice and food preparation techniques due to increased **multiculturalism**.
- Developments in scientific research and production of food have led to an increase in the production of **organic** and **genetically modified foods**.

Career opportunities and working conditions

Careers

Employment opportunities are spread across all food industry sectors and they vary in level from unskilled to semi-skilled to highly skilled. The following trends have emerged:

- The increased use of **mechanisation**, **automation** and **computerisation** in the industry has resulted in less use of manual labour.
- The competitive marketplace has spawned enormous growth in employment of **advertising and marketing** specialists.

- While Equal Employment Opportunities are part of any employment contract, it is apparent that some types of work in the food industry are more female (e.g. retail) or male dominated (e.g. transportation).

Examples of career opportunities in each food industry sector are as follows:

Fisheries and agriculture	Food processing and manufacture	Food retail	Food service and catering
Farmers	Technicians	Sales assistants	Chefs
Fishers	Fork-lift drivers	Packers	Wait staff
Boat builders	Production line workers	Display personnel	Bar staff
Fishing inspectors	Laboratory technicians	Butchers	Cleaners
Mechanics	Food technologists	Bakers	Kitchenhands
Scientists	Market researchers	Managers	Salespeople
Wholesalers	Computer technicians	Warehouse workers	Suppliers
Transport operators	Graphic designers	Cleaners	Delivery staff
Oyster farmers	Salespeople	Accountants	Accountants
Manufacturers of equipment for food production		Clerical staff	Artists (display boards and menus)
		Computer technicians	Advertising staff

Working conditions

The diverse range of employment opportunities creates greatly varied working conditions and employment awards. However, working conditions for food industry employees usually share the following features:

- Employees are frequently expected to work shift work and public holidays.
- Much of the work is casual or part-time employment.
- Payment varies depending on skill level and nature of work.
- Employment conditions and salary are frequently negotiated with a small business owner or covered by an award that is part of an enterprise agreement for union members.

Gender issues

Employing more than 315 000 people each year, the food, grocery and beverage industry offers a broad range of employment opportunities in a vast number of areas. The gap between men and women employed in the food industry is narrowing due to increased employment opportunities, more women entering full-time work, more women attaining tertiary qualifications and anti-discrimination laws in place. In saying this, women are still earning 83 cents for every dollar men earn, a 2009 survey by the Equal Opportunity for Women in the Workplace Agency (EOWA) has revealed.

Activity 5 **A PAGE 32**

Explain career opportunities and working conditions, including gender issues, within the Australian food industry.

1.3 Policy and legislation

Food policies are developed by the Australian and New Zealand Ministerial Forum on Food Regulations. The policies are a **framework** for the development of food standards.

Legislation is **law** passed by government that describes what can and cannot be done in specific situations. In the food industry, food legislation operates over three government levels—local, state and federal. An example of a federal law that relates to food is the *Trade Practices Act 1974*.

Who formulates and enforces food policy and legislation?

Governments are advised by independent organisations, known as **advisory groups**, on the development of policies and legislation. Advisory groups can be in the form of:

- **Business groups** within a food sector, such as the Australian Dairy Corporation and the Australian Food and Grocery Council
- A group that advises on specific **health** issues, such as the National Heart Foundation
- An independent body with the power to provide advice and make recommendations to government agencies, such as **Food Standards Australia New Zealand (FSANZ)** (see below)
- Groups that protect the food supply from **contamination**, such as the Department of Agriculture and Water Resources.

Food Standards Australia New Zealand (FSANZ)

Food Standards Australia and New Zealand (FSANZ) is an **independent legislative body** established by the *Food Standards Australia New Zealand Act 1991*. FSANZ's role in the food industry is to coordinate the following tasks:

- **Enforcement** and **revision** of food standards codes
- Development of **risk assessment** policies for imported foods
- Provision of **safety education** in food and food manufacture

- **Surveillance** of food available in Australia
- Development of codes of practice and food **product recalls** in order to ensure food safety.

Food safety is in fact **everybody's** responsibility. There are 11 500 cases of food poisoning daily in Australia and 120 deaths from food poisoning in Australia each year. Business owners, Food safety supervisors and food handlers are an integral part of ensuring food safety.

Australian Food Standards Code

The **Australian Food Standards Code** is the tool used by FSANZ to outline a general list of standards that food producers must follow, including:

- Food labeling (including use-by dates, additive and nutritional information)
- Standards affecting classes of foods such as cereals, eggs, meat, fish, fruit and vegetables
- Food safety standards that relate to residues and foreign objects in food.

Product recall

FSANZ has overall responsibility for **food recalls**. Each business within the food industry must have a product recall management plan in place.

Codes of practice

FSANZ develops codes of practice which are non-binding agreements that businesses are encouraged to follow. These codes of practice allow consumers to make **informed choices**. For example, if a manufacturer claims their product is 'fat reduced' it must contain at least 25% less fat than the regular product to which it is being compared and at least 3 g less fat per 100 g of food.

Government policies and legislation (local, state and federal) that impact on the Australian food industry

The two main policy areas that affect the Australian food industry are:

- National nutrition policies
- Trade policy.

National Nutrition Policy

In **2011** the Legislative and Governance Forum on Food Regulation agreed to develop a comprehensive **National Nutrition Policy**. The Nutrition Policy provides an overarching framework to identify, prioritise, drive and monitor nutrition initiatives within the context of the government's preventive health agendas.

The National Nutrition Committee developed the National Nutrition Policy and is chaired by the Department of Health and Ageing. The Committee also includes senior-level representatives of health departments from all states and territories, as well as the Department of Agriculture and Fisheries.

This policy provides strategies and direction on the following public health and nutrition issues:

- Prevention of overweight and obesity
- Increasing the consumption of vegetables and fruit
- Promotion of optimal nutrition for women, infants and children
- Improving nutrition for vulnerable groups.

A second key policy formed was the National Aboriginal and Torres Strait Islander *Nutrition Strategy and Action Plan 2000–2010 (NATSINSAP)*. NATSINSAP highlights seven key areas for action:

- Food supply in remote and rural communities
- Food security and socioeconomic status
- Family focused nutrition promotion: resourcing programs, disseminating and communicating 'good practice'
- Nutrition issues in urban areas
- The environment and household infrastructure
- Aboriginal and Torres Strait Islander nutrition workforce
- National food and nutrition information systems.

Trade Policy

Exports are the movement of food products out of Australia. This benefits Australia by increasing the global market share of profits that come back to Australia.

Imports are the movement of food products into Australia from other countries. This results in increased competition and increased product variety.

Australia has commitments under the **World Trade Organization (WTO)** on tariffs, quotas, export subsidies and domestic support for agricultural products.

Free trade agreements (FTAs) are treaties between two or more countries that benefit Australian importers, exporters, producers and investors by reducing and eliminating certain barriers to international trade and investment. Australia participates in FTAs with individual and groups of countries.

The Australian Government's policies aim to protect domestic industries, consumers and the environment from harmful and dangerous goods imported from overseas. In reality the removal of trade barriers has resulted in a large increase in imported foods.

The tools used by government to remove trade barriers include:

- Reducing **tariffs**
- Raising **quotas**
- Eliminating **subsidies**
- Establishing free trade agreements with other countries.

Activity 6 A PAGE 32

Read the article and answer the questions:

1. (a) Michel's Patisserie is part of a larger retail group. Name this group.
   (b) Identify other food retailers that belong to this group.
2. Michel's Patisserie has behaved illegally in relation to the sale of some of its products. Name this illegal behaviour.
3. Franchise owners of Michel's Patisserie are complaining about instructions from their supplier and management regarding the sale of their food products. List their concerns.
4. Identify the authorities that should be involved in the investigation of Michel's Patisserie.
5. What legislation has probably been breached?

Michel's expiry date extensions take the cake

Furious franchisees have been told to stretch use by dates by up to six months, writes Adele Ferguson.

National bakery chain Michel's Patisserie faces a food safety investigation after deliberately selling batches of chocolate cakes, birthday cakes and edible decorations to customers months after their use by date.

In a series of memos, obtained by the Herald, the chain's owner, Retail Food Group instructed Michel's franchisees to ignore expiry dates on packaging and adopt a new shelf life extension date, ranging between two months and six months.

The move has caused an uproar among franchisees, who are forced to buy RFG products under the terms of their agreements. Most of the products are delivered frozen and then they are thawed out and sold to customers.

Franchisees complain that sales have tumbled in recent years as the price of products increased and the quality fell when RFG moved to a frozen food model from freshly baked cakes, rolls and pies.

In one memo franchisees were told that coloured edible plaques with a use by date of January 15 this year had been extended to July 15.

"If you receive coloured plaques from this batch number that still denotes the original January expiry date, please disregard this and ensure staff are aware of the new expiry date," the memo says.

Other memos extend the use by date by three months on chocolate cakes, including tortes and birthday cakes.

Products including vegetable, and spinach and feta scrolls had their best before date extended by two months on the original packaging.

The extensions have triggered the NSW Food Authority to refer RFG for investigation to its Queensland counterpart, where RFG is headquartered.

A separate body that sets food standards, Food Standards Australia New Zealand, told the Herald that food should not be eaten after the use by date for health and safety reasons.

It said it was illegal to sell a product after the use by date, adding that while it was legal to sell foods that have passed a best before date, the food may have lost some quality. RFG refused to answer questions about the issue but said in a statement that any date extension was only considered after consultation with suppliers.

"RFG takes its food standard obligations extremely seriously and any potential breach of such standards would be investigated thoroughly alongside our suppliers," the statement said.

RFG said during the week it had met with franchisees to discuss ways to improve the network including how RFG would source, develop and deliver products to franchisees.

"RFG acknowledges that there have been challenges in the Michel's network and is actively working to improve these," it said.

One of the supplier's listed on the memos that calls on franchisees to extend the use by date of cakes - Homebush Cakes - failed to respond to questions.

A spokesperson for the NSW Food Authority said it has commenced an investigation into Homebush Cakes concerning the extension of use by dates.

It is the latest scandal to hit RFG, which was singled out for a special mention in a damning parliamentary report into the $170 billion franchise sector.

The March report said RFG's success relied on extracting profits from its franchise systems with hugely deleterious results for franchisees, which operate under the Michel's, Gloria Jeans, Pizza Capers, Brumby's and Donut King banners.

It recommended ASIC, the ACCC and the Australian Taxation Office investigate behaviour including possible insider trading, breaches of market-disclosure obligations and tax avoidance.

Wayne Hong, a former Michel's franchisee who closed his store last year and suffered a massive financial loss, said he lodged many complaints to RFG about the poor quality of products. He said cakes regularly arrived at stores damaged, chocolate ganache cakes were cracked and the taste of some cakes was bad. "It was really dodgy," he said.

A message sent to a franchisee Whatsapp group about the issue resulted in a series of photos of cakes and other products taken in the past week after being delivered with human hairs stuck in the icing, sausage rolls broken or over cooked, one cake frozen and refrozen, split pies and chocolate cakes with icing stuck to the box, indicating problems with the quality of delivery.

Insiders claim the expiry time shift was an act of desperation by RFG to avoid writing off old stock which would have an adverse impact on its finances.

One said RFG had thrown out hundreds of thousands of dollars worth of expired stock in recent years. The insider said RFG orders products from suppliers at agreed quantities and prices. The products are frozen and sent to distribution centres where they are sent to franchisees after an order is placed. "They are now just extending the shelf life and forcing us to sell expired product, which is preposterous for a company that deals with food. I was going to ask a government agency about the legalities of this but was concerned about repercussions," one franchisee said.

Another franchisee said store owners were being forced to buy old cakes and rolls at full price and each product sold attracted a royalty fee and a marketing levy.

"RFG could try to do right by their franchisees and reduce the cost of goods to entice us to purchase stock nearing its expiration date, but instead they don't discount the product and force us to sell out of date stock. We cannot return the products, as the distributor refuses to issue a credit given the shelf life extension," the franchisee said.

In one memo, RFG directs franchisees to extend the use by date on cakes including shimmer tortes and an RFG happy birthday cake from April to July 2.

A memo sent on March 15 says oven roasted vegetable scrolls and spinach and feta scrolls with a best before date of April should be extended to June. The memo says "the extension is on the condition that the products have been stored at the correct frozen temperature and handled correctly".

Some memos say RFG has received a shelf life extension from the supplier based on their shelf life testing protocols. Others have qualifications such as the extension is only valid if the cakes have been kept at or below minus 15 degrees stated on the supplier's specification for the duration of the shelf life to date.

A franchisee who leaked the memos said: "Ultimately we just want to be able to operate our business with pride in our product and our brand, and be able to make a comfortable living for the hard work we do week in and week out. But we have no faith or loyalty to RFG, it has been burnt out of us, so we are happy to help in any way to have more of their immoral conduct come to light."

Legislation

Food industry legislation is administered on all three levels of government. The legislation is reviewed regularly and appropriate changes or amendments are made. Examples of legislation administered at each government level include:

Federal

- *Food Standards Australia New Zealand Act 1991*
- *Imported Food Control Act 2018*
- *Competition and Consumer Act 2011*
- *Biosecurity Act 2015*
- *Gene Technology Act 2016*
- *Dairy produce Act 1986* amended 2017
- *Fisheries Management Act 1991*

State

Work Health and Safety Act 2011

Public Health Act 2017

NSW Food Act 2003

Fair Trading Act 1987

Fisheries Management Act 1994 amended 2017

Fertilisers Act 1985.

The Department of Primary Produce in NSW plays an active role in ensuring Federal government legislation is appropriately applied at State level.

Local

Local governments establish structures and roles to ensure the effective implementation of Federal government legislation. These include:

- Appointment of Environmental Health Officers (EHOs)
- Codes for inspection of food and food premises
- Codes for construction and alteration of food premises.

Legislative requirements for packaging and labelling

Manufacturers have obligations under a number of legislative requirements:

Under the National Food Standards Agreement, the National Health and Medical Research Council's Food Standards Code is the basis for labelling standards. The **Food Standards Code** sets labelling standards to ensure consumers are provided with correct information and safe food through proper packaging.

Other Acts include:

- The State ***Trade Measurement Act*** enforces correct labelling of weight of food.
- The ***Trade Practices Act*** ensures that imported foods have the correct country of origin on the label and prohibits misleading and deceptive conduct in respect to labelling and packaging.

National Packaging Covenant

This was established by the Australian New Zealand Environment and Conservation Council and the National Environment Protection Council in 1999. The covenant is designed to:

- Reduce the environmental impacts stemming from the disposal of used packaging
- Conserve resources through improved design and production processes
- Facilitate the re-use and recycling of used packaging materials.

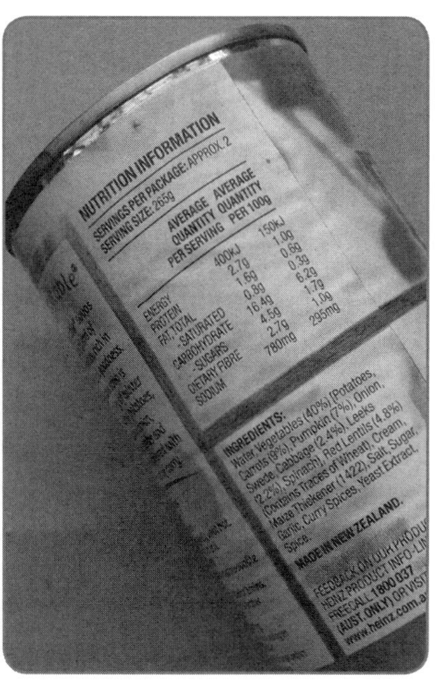

Producers are required to sign a covenant that demonstrates their undertaking to monitor all aspects of the packaging supply chain, from raw materials to kerbside recycling.

For more information on the National Packaging Covenant visit: www.packagingcovenant.org.au/.

Food labelling

Food labelling is controlled by the Australian Food Standards. It focuses on three areas:

- Statements or words that **must** appear on a label
- Statements or words that **must not** appear on a label

- Statements or words that **may** appear on a label under certain circumstances.

Labelling legislation requires that:

- Labels must be legible, of uniform size and in English.
- Foods must be labelled with an accurate name or description; for example, a 'strawberry' yoghurt must actually contain strawberries.
- Labels must have the name and business address in Australia or New Zealand of the supplier (manufacturer or importer).
- Where specific storage conditions are required in order for a product to keep until its 'best before' or 'use by' date, manufacturers must include this information on the label; for example, 'this yoghurt should be refrigerated'.
- Food additives must be listed in the ingredients list by its class name followed by the prescribed number or specific name in brackets; for example, thickener (414 or gum Arabic).
- In Australia there must be a country of origin statement.
- A food must be labelled as genetically modified if it has altered DNA or protein in the final product.
- A food that has been irradiated must be labelled as having been treated by irradiation.

In addition, in 2002 the Food Standards Code made amendments to the labelling of packaged foods:

- Mandatory nutrition information panels on nearly all packaged foods showing average energy content, protein, fat, saturated fat, carbohydrates and sugars, and sodium (salt).
- The major food allergens, such as eggs, milk, fish, peanuts and other nuts, sesame, soybeans, and cereals containing gluten (e.g. wheat) and their products, must be labelled, however small the amount.
- Date marking was clarified to 'use by' dates, which means the food cannot legally be sold or consumed after that date for public health and safety reasons, and 'best before' dates, where the food may still be sold after the date but may not retain some of the specific qualities for which any claims are made on the product.
- Ingredients must be listed in descending order (by ingoing weight).
- The percentage of the main or 'characterising' ingredient must be listed in the ingredients list; for example, the percentage of strawberry in a strawberry yoghurt.

Restrictions on labels

The following restrictions are placed on labelling:

- **No misleading** trade names.
- **No nutritional claims** that don't adhere to strict food code guidelines; for example, 'no added sugar' may be only used when the food or each ingredient contains no added sugars, honey, malt, malt extract or maltose.
- No claims of **therapeutic action**.
- No words, statements, claims or designs which could be interpreted as **advice of a medical nature**.
- The word '**pure**' may only be used with single-ingredient foods that contain no additives.
- The word '**health**' must not be used in the name of any food.

Activity 7 A PAGES 32–33

Choose one sector of the Australian food industry that you have studied and list in point form the main policies and legislation that are applicable to that sector. Briefly explain how they impact on that sector.

Chapter summary

The three areas of content that students should know are:

1. **Sectors of the Australian food industry**
   Sectors of the agri-food chain in the Australian food industry include:
   - Agriculture and fisheries
   - Food processing/manufacturing
   - Food retail
   - Food service and catering
   - Emerging technologies in food productioin, manufacturing and packaging.
2. **Aspects of the Australian food industry**
   The **operation** of organisations within the Australian food industry. Particular attention should be paid to:
   - Levels of operation and mechanisation, including household, small business, large companies, multinationals
   - Research and development
   - Quality assurance
   - Consumer influences
   - Impact on
     - the environment
     - the economy
     - society
   - Career opportunities and working conditions.
3. **Policy and legislation**
   Particular attention should be paid to:
   - Advisory groups that have a role in formulating and implementing policy and legislation
   - Government policies and legislation (local, state and federal) that impact on the Australian food industry.

Revision questions

Multiple choice   A PAGE 33

1. The food service and catering industry is growing rapidly because
   A the fresh food supply is unreliable.
   B people do not have the skills to prepare interesting meals.
   C Australians have busy lifestyles and little time to prepare food.
   D Australia's climate means fresh foods deteriorate quickly.

2. In recent years, the Australian food industry has responded in a number of ways to growing community concern for the environment. In which area has the response been greatest?
   A microwave processing
   B food storage
   C food packaging
   D factory waste disposal

3. The government legislations and policies which have had the most significant impact on Australia's food supply are the
   A *Trade Practices Act* and the National Nutrition Policy.
   B *Trade Measurement Act* and Trade Policy.
   C *Food Act* and Trade Policy.
   D *Occupational Health and Safety Act* and the National Nutrition Policy.

4. Food Standards Australia New Zealand is responsible for
   A the provision of safety education in food and food manufacturing.
   B regulating food prices.
   C imposing tariffs on all imported foods.
   D regulating the shelf life of all fresh foods.

5. The Australian Food Standards Code concentrates on which of the following aspects of food production?
   A reducing tariffs
   B quarantine inspections
   C the control of the use of additives
   D the control of smoke emissions during food production

6. The main aim of the National Nutrition Policy is to
   A educate people on wise food choices to reduce incidence of diet-related disease.
   B educate people on the benefits of breastfeeding.
   C reduce dangerous food production practices.
   D educate the public about the Australian Dietary Guidelines.

7. The increased use of value-added convenience foods has resulted from
   A increased multiculturalism.
   B more men preparing the meals.
   C increased number of women in full-time employment positions.
   D increased awareness of nutritional status of the Australian population.

8. Material minimisation is a strategy used by manufacturers to
   A reduce recycling.
   B increase landfill.
   C reduce the cost of the product.
   D reduce the weight of packaging but still maintain packaging strength.

9. The main energy source used in the manufacture of food is
   A wind.
   B coal.
   C water.
   D electricity.

Short-answer responses

A PAGES 34–36

10. (15 marks)
   (a) (i) Name ONE organisation within the Australian food industry. (1 mark)
   (ii) Identify the sector of the industry to which this organisation belongs. (1 mark)
   (b) Explain the activities of this organisation under the following headings:
   (i) Level of operation (2 marks)
   (ii) Research and development (2 marks)
   (iii) Quality assurance (2 marks)
   (iv) Consumer influences (2 marks)
   (v) Impact on the environment, economy and society (3 marks)
   (vi) Career opportunities and working conditions. (2 marks)

11. (15 marks)
   (a) Select ONE sector in the Australian food industry and describe TWO recent developments in that sector. (4 marks)
   (b) Analyse HOW policies OR legislation influence the operation of the sector which you have chosen. Use the following headings to assist your analysis:
   (i) Quality assurance (3 marks)
   (ii) Impact on the environment, economy and society (3 marks)
   (iii) Careers and working conditions (3 marks)
   (iv) Consumer influences. (2 marks)

12. (15 marks)
   (a) What is an advisory group? (2 marks)
   (b) Outline two advisory groups roles in formulating policy relevant to the Australian food industry (6 marks)
   (c) Discuss the difference between policy and legislation. (4 marks)
   (d) Name one legislation that is enforced from each government level of (3 marks)
   (i) Federal
   (ii) State
   (iii) Local.

13. (15 marks)

Analyse two environmental issues related to packaging and explain how packaging innovations in the package you have chosen are addressing these issues.

Structured extended response A PAGE 36

'Food manufacturing technologies impact on society.'

Discuss this statement in relation to environmental and social issues.

Answers Australian Food Industry

Activity 1

Term	Meaning
Agriculture	The cultivation of land to produce crops and/or animals
Agri-food chain	The production and supply of food for the consumer
AQIS (Australian Quarantine Inspection Service)	A government body whose main role is to protect Australia's food supply from contamination
Aquaculture	The production of plants or animals in water
Biotechnology	Use of microorganisms, or their products, for specific purposes such as the making of food with certain desirable attributes
Catering	The provision of food and service for functions, usually on a commercial scale
'Clean green'	A description of food which is produced in an environment free from chemical contaminants
Embargo	A ban placed on the importation of a particular product
Fisheries	The cultivation of various cold-blooded aquatic species, usually for a commercial or scientific purpose
FSANZ (Food Standards Australia New Zealand)	An independent legislative body whose main role is to standardise food laws
Genetic engineering	A process which involves removing a gene from a living organism and transferring the gene to another living organism. This is usually done to improve the characteristics of the organism which has the gene added, such as a quick-growing, high-yielding tomato plant
Legislation	A law passed by parliament that says what can and can't be done
Manufactured foods	Foods which have been made or produced by hand or, more frequently, machinery. Manufactured foods are often processed foods that have undergone a number of processing steps
Multinational	Belonging to more than one country, or influenced by more than one country
Policies	Government strategies put in place to achieve specific improvements, such as government policy on food safety issues, 'Food Safety Campaign Group'
Primary industries	Those businesses that produce food in its most simple form (i.e. plants and animals)
Primary produce	Food which is produced in its natural state, such as meat, fruit, vegetables, eggs
Processing	A series of actions that change food from its natural state to a slightly or significantly new form; for example, harvested, washed potatoes have undergone minimum processing whereas frozen chips are a more processed form of potato
Quota	The total amount of goods that can be imported

Recall	Removal of goods that may be a consumer safety hazard from sale, distribution and consumption
Subsidy	Direct aid given to food producers
Tariff	A duty or tax imposed by a government on imports or exports
Technology	The scientific approach used to improve productivity, quality and consistency (may be simple, such as a hand mixer, or complex, such as the process of monitoring supply of stock)
Value-added	The processing of goods to increase their selling price

Activity 2

Development	Sector/s	Explanation
One-stop shopping	■ Food retail	Supermarket shopping, loss of specialist retailers like butchers and greengrocers, and a reduction in small businesses
Irrigation	■ Agriculture and fisheries	Greater control over environmental conditions, automatic watering from water reserves increases yields and reduces crop losses due to harsh environmental factors
Convenience meals	■ Food processing and manufacture	Food service and catering Meals which require little or no preparation
Aquaculture	■ Agriculture and fisheries	Artificial breeding programs of water-borne animals or organisms
Home delivery service	■ Food retail ■ Food service and catering	Purchased items delivered to the purchaser's door (usually for a fee)
Genetic engineering	■ Agriculture and fisheries	Transfer of genes from living organisms to develop enhanced qualities in plants and animals
Value-added foods rather than primary produce	■ Food processing and manufacture	Trend towards manufactured foods rather than primary produce, increased profit margins, higher levels of technology, increased employment potential
internet shopping	■ Food retail	Online shopping, home deliveries without physically visiting a retail outlet
Breeding of lean meats	■ Agriculture and fisheries	Genetic engineering or feeding/breeding programs which reduce fat deposits in animals
Product scanning	■ Food retail ■ Food processing and manufacture	Pricing and stock control with scanning technology to provide readings of unit pricing and available stock
Pollution control	■ Agriculture and fisheries ■ Food processing and manufacturing ■ Food retail ■ Food service and catering	Controls of cleanliness of air, water, noise, waste management, use of chemicals in farming
Australia's 'clean green' image	■ Agriculture and fisheries	Reduction of chemical residue in plants and animals for human consumption
Production of low-fat animal products	■ Food processing and manufacture	Removal of fat or manufacturing procedures which reduce fat in animal products

Activity 3

Sector	Foods/meals that reflect that sector
Agriculture and fisheries	Primary produce such as fruit, vegetables, meat and dairy products
Food processing and manufacture	Canned goods Convenience foods
Food retail	Grocery lines
Food service and catering	Restaurant cuisine Food courts (i.e. takeaway foods)

Activity 4

(a) **Emerging technology: nanotechnology sector, food manufacture**

Nanotechnology focuses on the physical/biological structures smaller than 100 nanometres which result in unique properties because of their nanosize. Nanotechnology is beginning to be utilised in the food industry in the following ways:

- The study of nanostructures in biological materials of plant and animal origin could result in the development of nanomachines. These could be used to circulate through the blood stream and clean out fat deposits from arteries. They could also kill microbes and could be delivered to the human body through foods.
- Development of foods by non-biological means using advanced nanotechnology could be another future development, meant to ensure enough nutrition with limited resources.

(b)

Risks	Benefits
Emerging technology: genetic engineering	
■ **Imprecise technology**—A gene can be cut precisely from the DNA of an organism, but the insertion into the DNA of the target organism is basically random. As a consequence, there is a risk that it may disrupt the functioning of other genes essential to the life of that organism. ■ **Side effects**—Scientists do not yet understand living systems completely enough to perform DNA surgery without creating mutations which could be harmful to the environment and our health. ■ **Widespread crop failure**—Genetic engineers intend to profit by patenting genetically engineered seeds. This means that when a farmer plants genetically engineered seeds, all the seeds have an identical genetic structure. As a result, if a fungus, virus or a pest develops which can attack this particular crop, there could be widespread crop failure. ■ **Threatens our entire food supply**—Insects, birds and wind can carry genetically altered seeds into neighbouring fields and beyond. Pollen from transgenic plants can cross-pollinate with genetically natural crops and wild relatives.	■ Genetic engineering reduces the costs of production, meaning that the poor can afford more food, and be more self-sufficient. ■ Better resistance to weeds, pests and disease ■ Better texture, flavour and nutritional value ■ Longer shelf life, easier shipment ■ Better yield, more efficient use of land ■ Less herbicides and other chemicals ■ Cheaper and safer source of human medicine ■ Pest and disease resistant ■ Herbicide resistant ■ Easier to harvest plants

Risks	Benefits
Emerging technology: genetic engineering (continued)	
■ **No long-term safety testing**—Genetic engineering uses material from organisms that have never been part of the human food supply to change the fundamental nature of the food we eat. Without long-term testing no one knows if these foods are safe. ■ **Toxins**—Genetic engineering can cause unexpected mutations in an organism, which can create new and higher levels of toxins in foods. ■ **Allergic reactions**—Genetic engineering can also produce unforeseen and unknown allergens in foods. ■ **Decreased nutritional value**—Transgenic foods may mislead consumers with counterfeit freshness. A luscious-looking, bright red genetically engineered tomato could be several weeks old and of little nutritional worth. ■ **Increased use of herbicides**—Scientists estimate that plants genetically engineered to be herbicide-resistant will greatly increase the amount of herbicide use. Farmers, knowing that their crops can tolerate the herbicides, will use them more liberally.	■ Increased productivity ■ Fewer agricultural and horticultural chemicals needed ■ Fewer chemical residues in food chain ■ Remove undesirable characteristics ■ Design what people want or need ■ Improve or add desirable characteristics ■ Reduce management costs
■ **Ecology may be damaged**—The influence of a genetically engineered organism on the food chain may damage the local ecology. The new organism may compete successfully with wild relatives, causing unforeseen changes in the environment.	
■ **Gene pollution cannot be cleaned up**—Once genetically engineered organisms, bacteria and viruses are released into the environment, it is impossible to contain or recall them. Unlike chemical or nuclear contamination, negative effects are irreversible.	
Emerging technology: Genomics	
■ Consumer acceptability could be an issue.	■ Can gene map for optimum breeding and meat quality ■ Reduce the number of cross trials, saving time and money
Emerging technology: Transgenic meats	
■ Consumer acceptability could be an issue ■ Danger that if transgenic animals escaped they could out-breed traditional stocks causing imbalance in food supply for these animals	■ Can reduce effects of chemicals on the environment ■ Injection of growth hormones can result in leaner meat and the ability of the meat industry to have stock reach market potential quicker, so improving profit and supply.
Emerging technology: Packaging technology	
■ New technologies could impact on cost if new machinery, equipment and so on is required	■ Material minimisation results in reduced weight of packaging but still maintains packaging strength ■ Reduced landfill ■ Improved recycling practices
Emerging technology: Organic farming	
■ Foods can be more expensive ■ Lower productivity	■ Chemical-free production ■ Environmentally friendly agricultural practices mean less harm to the environment

Activity 5

- Approximately 191 400 people were employed in food and beverage manufacturing in Australia in 2006–07.
- All sectors of the Australian food industry have diverse employment opportunities. Trends such as improved mechanisation and computerisation have resulted in less use of manual labour.
- Relating to gender, some aspects of the Australian food industry are female-dominated, such as food retail, while in manufacturing and processing the transport aspect of food manufacture is dominated by males.
- Working conditions in each sector are varied and diverse. Common features of employment in the Australian food industry include use of shift work and public holiday work. Food service, catering and food manufacturing are sector examples.
- Work is often casual or permanent part-time.
- Employment conditions and salary are often set by awards set as a part of an enterprise agreement for union members.

Activity 6

1. (a) Retail Food Group

   (b) Gloria Jean, Pizza Capers, Brumby's and Donut King

2. It is illegal to sell a food product after the use-by date.

3. Franchisee concerns are as follows:

- Instruction to sell food after the use-by date
- Instruction to sell food after the best-before date
- Food is delivered frozen and later defrosted.
- Food quality is not of a high standard.
- Food is often damaged.
- Food prices are not reduced if they have a short shelf life.
- Delivery problems
- Food cannot be returned.

4. Food Standards Australia and New Zealand; The Australian Competition and Consumer Commission; NSW Food Authority; Environmental Health Officers

5. *Competition and Consumer Act;* Food Standards Code; *Work Health and Safety Act; NSW Food Act; Public Health Act*

Activity 7

Sector example: Agriculture and Fisheries

- Trade Policy—role of World Trade Organization; Free Trade Agreements. Discuss how the Policy works and the advantages of its application—removal of trade barriers resulting in increased competition in the marketplace so advantages occur in price, competition and product variety. Role of subsidies, tariffs and embargoes.

- *Food Standards Australia and New Zealand Act*
- *Competition and Consumer Act*
- *Imported Food Control Act*
- *Fisheries Management Act*
- *Dairy Produce Act*
- *Gene Technology Act*
- *Public Health Act*
- *Fertilisers Act*
- *Fair Trading Act*

Multiple choice

1. **C** Australian lifestyles have changed significantly: women are working longer hours and less time is allocated to meal preparation in the home (one in four meals are eaten away from home).
2. **C** Developments in food packaging design and materials reflect awareness of:
   - Reduced pollution due to emissions in production of packaging materials
   - Reduced landfill due to increased reusability and recylcability
   - Awareness of solvents used in labelling
   - Reduction of packaging weight by making materials stronger and thinner, resulting in reduced transport pollution as more can be transported at once.
3. **C** The *Food Act* enforces food standards and hygiene regulations in the food industry and so deals with more levels. The Trade Policy and 'free trade' have increased food trade by removing trade barriers. The other choices are more specific or do not impact on supply.
4. **A** FSANZ is responsible for the provision of safety education in food and food manufacturing in the following areas:
   - **Revision** and **enforcement** of the Food Standards Code
   - Development of **risk assessment** policies for imported foods
   - **Surveillance** of food available in Australia
   - Development of codes of practice and food **product recalls** in order to ensure food safety.
5. **C** The Australian Food Standards Code controls the use of additives. Tariffs are controlled by the Federal Government under Trade Policy. Quarantine inspections are controlled by the Department of Agriculture and Water Resources. Protection of the retail sector comes under the *Competition and Consumer Act*.
6. **A** The main aim of the National Nutrition Policy is to educate people on wise food choices to reduce the incidence of diet-related disease. Dangerous food production practices are controlled by the ANZFA. Australian Dietary Guidlines and education on breastfeeding would be enforced by the Commonwealth and State health departments.
7. **C** Increased number of women in full-time positions. Traditionally women prepared most meals; however, with this increasing trend households are using more convenience foods due to lack of time but also increased expendable income.
8. **D** Reduce the weight of packaging but still maintain packaging strength. Material minimisations is a strategy introduced by manufacturers to reduce waste and cost by reducing the weight of packaging but still maintaining packaging strength.
9. **B** Coal. The main energy used in food manufacture is non-renewable resources (e.g. coal).

Short-answer responses

10. (a) (i) One **organisation**. Numerous answers, such as Pace Farms (agriculture and fisheries), Masterfoods (food manufacture), Coles (food retail), Pizza Hut (food service and catering) or any other organisation in any sector.

(ii) The **sector** depends on the organisation chosen in part (i). Any of the four sectors identified above would be accurate as long as it matches the organisation.

(b) (i) **Level of operation**. Depending on the organisation selected it can be household level, small business, large company or multinational. Influenced by number of people employed and where the organisation is based and operating.

(ii) **Research and development**. This depends on the level of operation, and generally larger businesses spend more on R & D. However, some smaller companies access research undertaken by agencies such as CSIRO or ANZFA. Candidates need to identify R & D specific to the organisation and who/how/why the research occurred.

(iii) **Quality assurance**. Candidates should identify the application of HACCP or other quality control procedures to the organisation they write about, such as consumer complaint lines, quality assurance guarantees, inspection programs or monitoring procedures.

(iv) **Consumer influences**. Depending on the organisation chosen students should explain the relevance of issues such as lifestyle, technology and skills, trends/fashions or marketing procedures within the sector or organisation.

(v) **Impact of the environment, economy and society.** Mention issues such as farming methods, waste management, transportation and storage; economic circumstances, company profit, taxation, subsidies and tariffs; lifestyle, family structure, health issues, choice/promotional strategies. These issues are all relevant and should be applied to the organisation and sector chosen.

(vi) **Career opportunities and working conditions**. Select a range of careers relevant to the organisation chosen, identify hours of work, salary structures, skills required, training/career advancement, issues such as clothing, grooming, trade unions, bonuses/incentives etc.

11. (a) (i) **Sector**: Either Agriculture and fisheries, Food manufacturing, Food retail or Food service and catering.

(ii) **Developments** will depend on the sector selected. It may include organic farming, genetic engineering, specific technologies relevant to the production of food items, packaging or processing equipment, methods of production or processing, internet shopping, scanning for pricing or stock control, home delivery services, take-away foods, ready-to-eat meals, breakfasts away from home etc.

(b) Policies OR legislation

(i) **Quality assurance**. HACCP, Occupational Health and Safety legislation, role of health surveillance officers (local government), AQIS (international trade), Codex Alimentarius, ANZFA, *NSW Food Act*.

(ii) **Impact on the environment, economy and society**. Clean air, clean water, noise pollution, labelling, use of additives (*Food Act*), *Trade Practices Act.*

(iii) **Careers and working conditions**. Award wages, working conditions, minimum wage, Occupational Health and Safety, workers' compensation.

(iv) **Consumer influences**. Building codes, *Trade Practices Act*, health and nutrition policy, Foods Standards Code, role of advisory groups.

12. (a) An advisory group is an independent organisation that guides or advises governments in the development of policies and legislation.

(b) (i) The National Farmers Federation provides advice to the government on issues affecting Australian farmers and agriculture productivity. Their motivation is to maximise competitiveness for Australian farmers. Their advice to the government might relate to farm business and productivity; natural resource management; biosecurity; health and welfare; education and training; or workplace relations.

(ii) Australian Food and Grocery Council. This council has over 170 member companies. It ensures that the Federal Government considers the views of the food industry when developing economic and industrial policies in such areas as reduction of trade barriers and tax concessions. Ultimately, such considerations will impact on the food industry in terms of product competition and pricing.

(c) Policies are strategies put in place by governments to improve all aspects of living. Legislation is a law passed by government that describes what can be done in the food industry. It is administered on a federal, state and local level. Give examples.

(d) (i) Federal legislation:

- *HACCP*
- *ANZFA Act*
- *Export Control Act*
- *Competition and Consumer Act*

(ii) State legislation:

- *Protection of the Environment Acts*
- *Food Act*
- *Work Health and Safety Act*
- *NSW Fair Trading Act*
- *Public Health Act*

(iii) Local legislation:

- Appointment of environmental health officer
- Codes of inspection (food premises)
- Codes of construction

13. Environmental issues include:

- Recycling
- Reusability
- Reducing landfill
- Reducing energy costs
- Less emissions in the atmosphere
- Material minimisation.

Students choose two of these environmental issues to discuss in relation to a form of packaging; for example, links between canning material minimisation and recycling.

Innovations in canning—link the following to these issues:

Material minimisation. Steel alloys used in can production are being developed so that cans can be produced with thinner walls, thus reducing costs and the use of energy resources, while still having the strength of traditional cans.

Production of plastic cans that can be recycled. Plastic cans are being produced that are capable of being heated to 120 °C and also provide gas barriers. This reduces the use of laminates or coatings used in conventional steel cans, therefore saving resources.

Structured extended response

Responses should show a clear interrelationship between technologies and their impact. In discussion, students should show how technologies are benefitting society in these two areas.

The easiest approach is to treat each issue separately and discuss how technologies in each issue impact in both positive and negative ways. The study guide notes address this adequately.

Environmental issues:

- Energy usage—The main energy used in food manufacturing is non-renewable resources such as coal. Food manufacturers need to investigate more efficient ways of using these resources so that society is not depleted of them. Ongoing extensive usage could result in exhaustion of non-renewable resources in years to come.
- Pollution—Technologies in food manufacturing need to be improved to reduce the effect of harmful production emissions on our atmosphere. For example, the types of materials produced also need consideration: gases released into the atmosphere during the production of such packaging products as glass contain more carbon dioxide than is released during the production of plastic packages, and plastic products are also recyclable.
- Waste disposal/recycling—Recycling and waste disposal methods need careful consideration so that the impact on society in terms of pollution, on-costs and use of resources is reduced. Recycling contributes advantages such as minimising landfill waste due to reuse of resources rather than use of new resources, hence saving future resources for society. Another example is the use of biodegradable packaging which is made from biodegradable material that will decompose through the action of micro-organisms in a reasonabie length of time. This will also contribute to reduced landfill and toxic emissions from decomposition.

Social issues:

- Lifestyle changes—Linked to an increase in convenience packaging and foods.
- Employment opportunities—Increased opportunities in food retail, food product development and food service and catering.

2 – Food Manufacture

Developments in food manufacture have an impact on society and the environment. A knowledge and understanding of food manufacturing processes informs choices and encourages responsible patterns of consumption.

Outcomes

On completing this chapter a student should be able to:

1. Explain manufacturing processes and technologies used in the production of food products (H1.1)
2. Apply principles of food preservation to extend the life of food and maintain safety (H4.2).

Activity 1 A PAGE 64

Following is a list of terms you should know as a result of completing this module.

Term	Meaning
Active packaging	
Automated	
Computerised	
Critical control points	
Enzymatic activity	
Equipment	
Food additive	
Food deterioration	
Food manufacture	
Food preservation	
HACCP	
Integral processes	
MAP packaging	
Microbial contamination	
Preservation processes	
Principles of food preservation	
Processed food	
Production run	
Quality assurance	
Quality control	
Raw material	
Sous vide	
Specification	

The HSC Examination

In the HSC Examination students could expect to answer questions on Food Manufacture to the value of approximately 25% of the paper. Questions may require students to integrate knowledge, understanding and skills developed through studying the entire course. It is therefore difficult to predict the weighting of marks in each section of the paper.

The paper may include:

- Multiple-choice questions
- Short-answer questions (which may include parts)
- A structured extended-response question, which includes two or three parts with one part worth at least 8 marks

OR

- An extended-response question (approximately 600 words).

2.1 Production and processing of food

Food manufacturing is the conversion of **raw materials** into a **final food product** using physical and chemical processes that must abide by safety, legal and environmental constraints.

Quality and quantity control of raw materials used for food processing

A **raw material** can be defined as any product that is used in the **manufacture** of another **processed food**.

Raw materials used in food processing are classed into four major areas:

1. **Ingredients** such as flour and margarine	
2. **Materials** used in the manufacturing process such as steam and water	
3. **Food additives** such as colours and flavours	
4. **Packaging** such as bottles and cans.	

Food manufacturers spend a large percentage of their **costs** on **raw materials**. Due to this, strict **specifications** are put in place to **reduce** the risk of **contaminated raw material** being used. These specifications are verified by the food manufacturer when raw materials are delivered.

Raw material specifications are established for each raw material used, according to the following criteria:

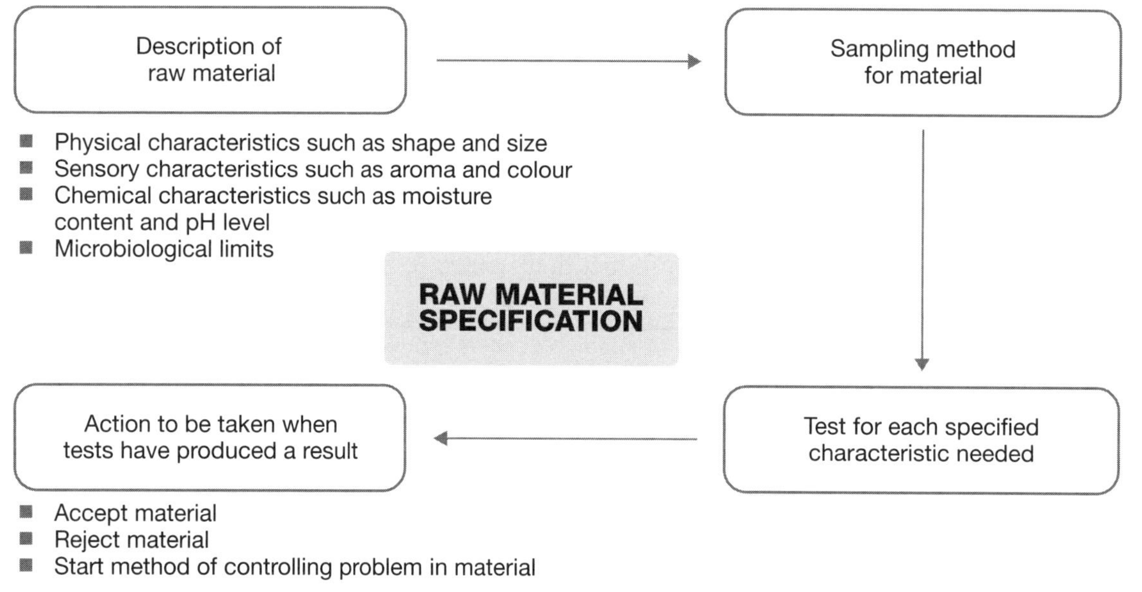

Figure 2.1 Raw material specifications

Food additives are substances added to food that are not normally consumed on their own.

The functions of food additives are to:

- Improve food **stability and shelf-life**
- Improve **sensory characteristics** of the food, that is, taste and appearance
- Adapt foods to provide for **special dietary needs**.

Students need to be aware of the types of additives and what they do in specific food products:

Additives	Function in specific food products
Flavour	Restore flavour and odour lost through food processing
Flavour enhancers	Improve the existing flavour of a product
Colours	Either restore colour lost in the processing or improve the physical appearance of the food
Vitamins/minerals	Replace nutrient loss that occurs during processing and storage
Mineral salts	Improve texture and mouthfeel of processed foods
Antioxidants	Prolong the shelf-life of a product by preventing oxidation
Humectants	Absorb moisture from the atmosphere so food does not dry out
Emulsifiers	Allow oil and water to mix in the production of foods using these ingredients
Preservatives	Inhibit the growth of bacteria, yeast, mould and viruses
Food acids	Maintain a constant level of acidity; produce a sharp taste
Thickeners	Make a food thicker
Artificial sweeteners	Impart a sweet taste in low-joule foods
Vegetable gums	Impart consistency and texture
Bleaching agents	Whiten foods

Code numbering of additives

Since 1986, code numbering has been used to identify **food additives** on food labels. This was an easier method to use because:

- Some additive names are so long they would take up too much space
- It brought Australia in line with the **international coding system**
- It simplified identification of which additive is present in a food product, especially for people with an **additive intolerance**.

ANZFA provide a book that has **all additive codes listed** and what they stand for. Food technology texts also provide this information.

INGREDIENTS_
Sugar, wheat flour, vegetable fats and oils [emulsifiers (471, 477), antioxidant (320)], raising agents (450, sodium bicarbonate), tapioca, starch, thickener (1442 from maize), salt, emulsifier (481), colour (E102), natural flavour

Contains wheat.
May be present: milk and tree nuts

Made in Australia from local and imported ingredients.

Activity 2 A PAGE 65

Take two different food products from a food cupboard and list all ingredients in that product. Identify:

(a) The ingredients that are **additives**

(b) The **function** of each additive.

An example is outlined for you with the additives underlined:

Chocolate dessert mix ingredients:

sugar, <u>thickener (starch)</u>, <u>mineral salts (450,339)</u>, cocoa, <u>food acid (263)</u>, vegetable fat, <u>vegetable gum (412)</u>, <u>colour (150)</u>, salt, flavour.

Characteristics of equipment used in different types of production and the factors influencing their selection

Unit operations are the specific processes food undergoes during production.

Processes are:

- Separation (physical and chemical)
- Grinding and milling
- Mixing
- Heating
- Cooling
- Freezing
- Evaporation
- Dehydration.

The following table is a description of processing techniques, equipment, storage and distribution systems used in industry:

Process used	Purpose	Domestic equipment used	Industrial equipment used	Food example
1. Separation				
(a) Filtration	The process of passing a liquid through a filter so solid particles are removed	Sieve	■ Filtering systems ■ Microbial air filters	■ Separation of cheese curd from water ■ Used in dryers in the production of powdered products (e.g. milk)
(b) Sedimentation	The use of gravitational or centrifugal forces to remove solids from liquids		■ Centrifuge ■ Spray drier	■ Separation of cream from whole milk
(c) Centrifuging	The process used to separate food particles of different densities		■ Centrifuge	
2. Grinding and milling	The process of reducing raw material size which may be required to: ■ Make raw materials easier to handle ■ Make raw materials suitable for the final product ■ Make a completely new product.	■ Food processor ■ Grinder ■ Mortar and pestle	■ Computerised mills	■ Grinding salt so easier to handle ■ Grinding peanuts to make peanut butter ■ Flour from wheat
3. Mixing	Process used to evenly distribute ingredients through a product batch	■ Electric beaters ■ Wooden spoon ■ Food processor ■ Bamix ■ Blender	■ Defoaming mix equipment ■ High-pressure mixer ■ Large-scale mixing equipment ■ Computerised mixing equipment	■ Equipment used for mixing is dependent on purpose and what is being mixed (e.g. a mixer for liquid has a different arm than a mixer for dough due to material thickness)
4. Heating equipment	Process used to heat raw materials during processing	■ Stove ■ Oven ■ Microwave oven	■ Industrial ovens ■ Heat exchangers: tubular and plate heat exchangers	■ Type of equipment used depends on the nature of the raw materials and intensity of heat treatment; for example, tubular heat exchangers are used for sauces, while plate heat exchangers are only suitable for liquids such as fruit juices

Process used	Purpose	Domestic equipment used	Industrial equipment used	Food example
5. **Cooling**	Process used to reduce temperature of a product to slow down or stop the activity of micro-organism and enzymes	■ Refrigerator	■ Heat exchangers where steam is replaced by chilled water and refrigerants ■ Industrial refrigerator	■ Fruit and vegetables
6. **Freezing**	Process used to change water content to ice, rendering it unavailable for microbial growth or enzyme activity	■ Freezer	■ Air blast freezers ■ Plate freezers ■ Immersion freezers	■ Meat and cakes ■ Fish and meat ■ Fresh vegetables and ice-cream
7. **Evaporation**	Process used to increase the solids concentration of a liquid food—liquid in the food is changed to steam or vapour	■ Oven	■ Industrial evaporator	■ Conversion of tomato juice to tomato paste
8. **Dehydration**	Process used to reduce the moisture content of the product to a level that limits microbial growth.	■ Drier ■ Salting	■ Cabinet dryer ■ Tunnel dryers ■ Bin dryers ■ Vacuum dryers ■ Drum dryers ■ Freeze drying ■ Spray drying	■ Fruits ■ Coffee ■ Tea ■ Herbs ■ Powdered milk

All basic **food production**, either **industrial or domestic**, uses one or more of the processes outlined above. Students need to know the **functions** of these processes and examples of **equipment** that carry out these processes in a domestic and industrial setting.

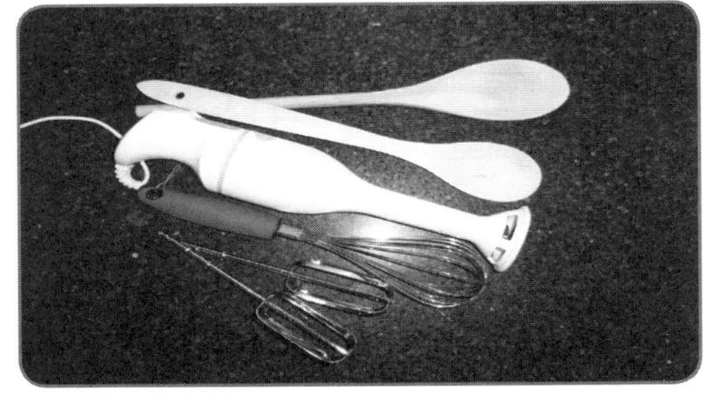

Activity 3 A PAGE 65

From a class study of the production of a food product, outline the **integral processes**.
List the equipment that would be used in:

(a) An industrial setting

(b) A domestic setting.

If you have not done this in class, research bread-making as information on the production of bread is readily available. The example in the answer section is ice-cream.

Production systems

> Production systems are the ways the processes of food production are **organised and applied**.

Production systems are classed as:

- Large-scale
- Small-scale
- Manual
- Automated
- Computerised.

Production systems generally follow this format, depending on the scale of the operation:

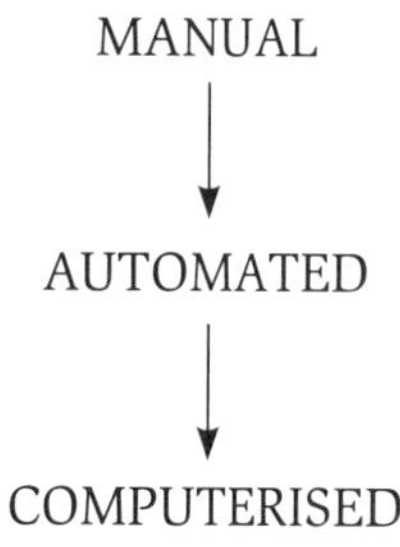

Large-scale production systems

These are used within **larger production operations**. Such systems are **based on a production line** which is a continual process from the inspection of **raw materials**, through to the storage and distribution of the final product. A large-scale system has the **advantage** of producing **large volumes of product** in a **short time** period.

Small-scale production systems

These systems are used on a **domestic level** and are **smaller in scale** and **less complex** in operation.

Manual production

This involves the **operator physically adding food** components such as filling pie cases by hand.

Automation processing

This is when **machines handle and control the processing** from raw materials to the finished product. Automated systems have set pre-determined conditions for such operations as sterilisation temperatures and times, humidity levels and mixing times.

Computerisation

In automation, **computerisation** is achieved through the **use of sensors**. Sensors enable large and complex operations to be undertaken due to data being stored and compared to pre-set specifications. Computer programs have been developed to respond to variations in raw materials and conditions. Cheese making is an example of food production being controlled by computerisation.

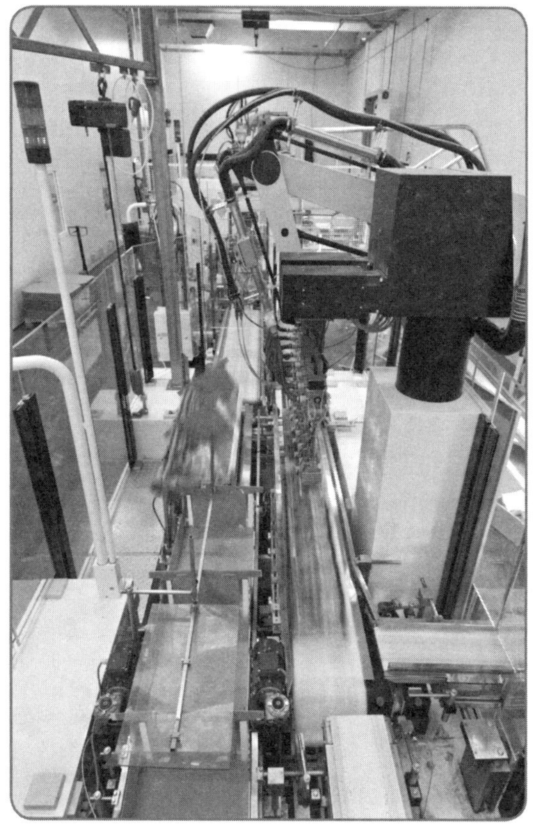

The type of production system used is dependent on the following factors:

- **Nature** of the product
- **Scale** of production, that is, amount required
- **Economic** consideration of which system would give the manufacturer a market and profit advantage
- **Consumer acceptance** of more impersonal approach to food service, that is, use of vendor machine instead of face-to-face selling.

Flow diagrams

A method of representing a food production process is through a flow diagram. Features of a flow diagram include:

- The use of **five basic symbols**
- Symbols are **written vertically** down the page with the symbol on the left-hand side, explanation on the right
- Symbols should be **joined by vertical lines**
- Symbols can be **used separately or joined**, depending on the process.

Production process flow chart

An example of a production process flow chart would be:

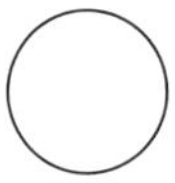

Operation
This is when a raw material is:

- Changed in some way
- Assembled or disassembled from another material
- Prepared for another operation.

Inspection
An inspection occurs when a material is checked, sampled, examined, measured or compared with a standard.

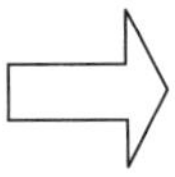

Transportation
This is when a material is moved from one place to another.

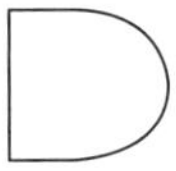

Delay
This is when conditions do not allow the next step of processing to take place.

Storage
This is when a product or raw material is kept in storage.

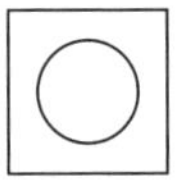

Combined symbols
Two symbols can be combined when two activities occur at the same time (e.g. heating and measuring temperature, which are examples of an operation and inspection combined).

Breakfast cereal flow chart

An example of a flow chart for a breakfast cereal would be:

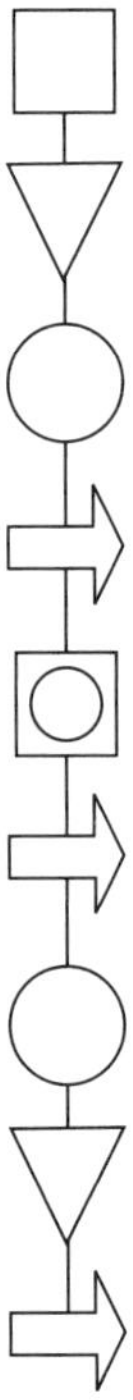

Wheat is inspected.

Wheat is stored below 20 °C.

Wheat's physical and chemical characteristics are changed due to heating.

Wheat travels in a steady measured stream to the mills.

Physical characteristics changed at this stage (moulded to form biscuit shapes) and inspected.

Biscuits are then dropped onto another conveyor belt.

Physical characteristics changed (colour) as shapes are toasted. The cereal is then packaged.

Cereal is stored.

The cereal is transported to the wholesaler for distribution.

Exam tip: Flow diagrams are a handy tool for summarising key production processes of a variety of food. The breakfast cereal flow chart is a good example of this.

Activity 4 A PAGE 65

Draw a flow chart for the following canned product, using the production methods outlined below. The methods are written in the correct order:

Quality assessment
Grading and sorting
Washing
Peeling
Blanching
Chilling and testing
Inspection
Can fill
Vacuum seal
Sterilise
Cool
Label
Package
Distribute

Quality management considerations in industrial practices to achieve safe food for public consumption

Quality management aims to control all aspects of production and produce a **quality food product** for the consumer. Quality management covers all aspects of:

- Product development
- Production
- Marketing
- Servicing.

Quality assurance is the **co-operation** of an organisation to achieve **quality control** of products. Quality assurance procedures include:

- A final product specification statement of the **level** of quality to be attained
- Methods for assessing and **measuring** the quality of the final product
- Clear **specifications** for all production areas, such as all processing areas to be free of contaminants
- Sampling and **testing** of the completed product.

> Food safety is a major quality assurance issue. Quality management strategies and occupational health and safety techniques are used to ensure safety of food for consumers and work conditions of food production workers.

Quality management techniques include **HACCP** and occupational health and safety.

Hazard Analysis of Critical Control Points (HACCP)

This is a quality management technique that identifies **potential hazards** within the production of a specific food product, and methods of dealing with them.

HACCP can be summarised as follows:

Assessment of hazards.

↓

Identification of critical control points (parts of the production process identified to be a point where a hazard can occur).

↓

Setting standards for each critical control point.

↓

Monitoring of critical control points.

↓

Description of clear procedures for prompt action if standards not met.

↓

Keeping accurate records to identify variations from the standards.

↓

Assessment of the system.

Occupational health and safety

Due to the complexity of food processing, manufacturers need to be aware of **potential accidents** for workers.

Employers in the food industry are required by law to:

- **Instruct** or train about safety issues
- Provide **well-maintained equipment** that functions efficiently when used correctly
- Provide a **safe environment** in line with the *Occupational Health and Safety Act 2000*. A safe environment includes adequate lighting, ventilation, washroom facilities, first aid provisions and correct on-the-job training.

If there are **20 or more employees** in a workplace, then an OH&S committee has to be formed. It is the responsibility of the committee to identify key accident areas, devise an accident report form and plan a course of action should an accident occur.

It is the responsibility of employees under **OH&S guidelines** to:

- Take care for the health and safety of others
- Wear correct protective clothing
- Use equipment as instructed
- Provide notification of accidents.

OH&S is the responsibility of all workers in the workplace.

Activity 5 A PAGE 66

(a) For a food production company you have studied, outline **quality control procedures** used in the production of a food product.

(b) Draw a flow chart for one of their products and **identify critical control points**.

2.2 Preservation

Food preservation is the processing of food to eliminate the conditions that cause spoilage.

The main reasons for preserving food are to:

- Keep food **safe** for human consumption
- Keep foods in a state **acceptable** for the consumer and so reduce waste
- Retain the **nutritive value** of the food
- Make perishable foods **available all year round** which provides for greater diet variety
- Achieve economic viability for food producers by reducing seasonal fluctuations in availability.

The causes of food deterioration and spoilage are:

Chemical reactions
Chemical breakdown of food is caused by chemical reactions within the food, such as rancidity of milk, or if the food becomes chemically contaminated from such sources as agricultural chemicals.

Physical reactions
Physical changes are those changes that alter the physical state of a food in some way. Examples of physical changes include size, surface damage or colour change. Causes of such changes can come from bruising, movement of food, freezing, burning and pressure.

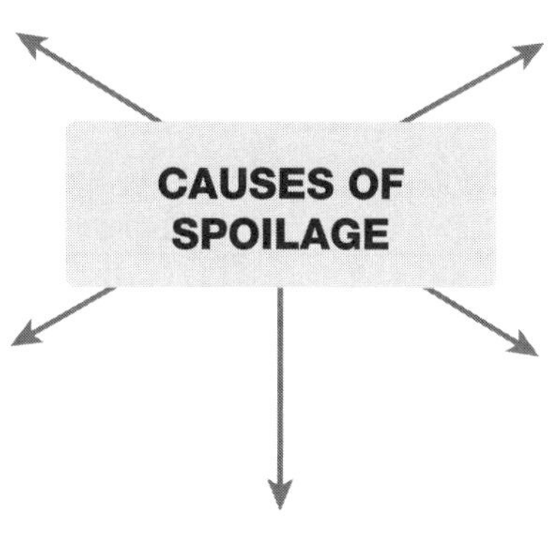

Microbial activity (i.e. bacteria, yeast, mould).
Microbial activity includes the effects of:

- bacteria
- mould
- viruses.

Given the correct conditions (i.e. food source, warmth and moisture) such micro-organisms will multiply and cause food spoilage.

Living organisms
Organisms such as insects and rodents can cause spoilage of food, such as weevils in flour and the transmittance of bacteria by rodents.

Enzymatic changes
Enzymes are found in meats, fruits and vegetables. Even when food sources are slaughtered or harvested, enzymes continue to work and so over-ripening can occur, causing food tissue decomposition.

Figure 2.2 Causes of food spoilage

The principles of food preservation refer to **control** of factors that allow types of **deterioration**. The factors that need to be controlled are:

- Restriction of water
- Addition of chemicals
- Temperature control
- Level of acidity or pH level
- Exclusion of air.

The following table outlines the food spoilage factor, preservation principle and examples of preservation methods:

Food spoilage factor	Preservation principle	Preservation method
Microbial activity ■ Bacteria ■ Yeast ■ Mould ■ Virus	**Temperature control** Micro-organisms grow or produce toxins in specific temperature ranges, thus the appropriate high or low temperatures are used	■ Pasteurisation ■ Sterilisation ■ Freezing ■ Chilling ■ Canning ■ Ultra heat treated (UHT)
	Restriction of moisture The amount of available water in a product will determine whether an organism can grow	■ Smoking ■ Drying ■ Freeze-drying ■ Salting ■ Sugar
	Acidity/pH level Most pathogenic microbes will not tolerate a pH of less than 4.2	■ Adding acid ■ Fermentation
	Exclusion of air ■ Most organisms need the presence of oxygen to grow ■ Not allowing food to come in contact with micro-organisms in aseptic conditions	■ Vacuum packaging ■ Aseptic packaging
Chemical reactions	■ Handling/storage n Agricultural practices	■ Correct storage (e.g. use of refrigeration) reduces the risk of chemical reactions such as rancidity ■ Using agricultural practices that reduce the risk of chemical residue
Enzyme activity	■ Exclusion of air ■ Handling ■ Temperature control ■ Acidity/pH	■ Reduction in oxygen levels and temperature can slow enzyme activity ■ Correct handling ensures minimal bruising so that oxygen makes less contact with internal tissue ■ Use of acid, such as citric acid, to reduce browning caused by enzyme activity

Preservation processes

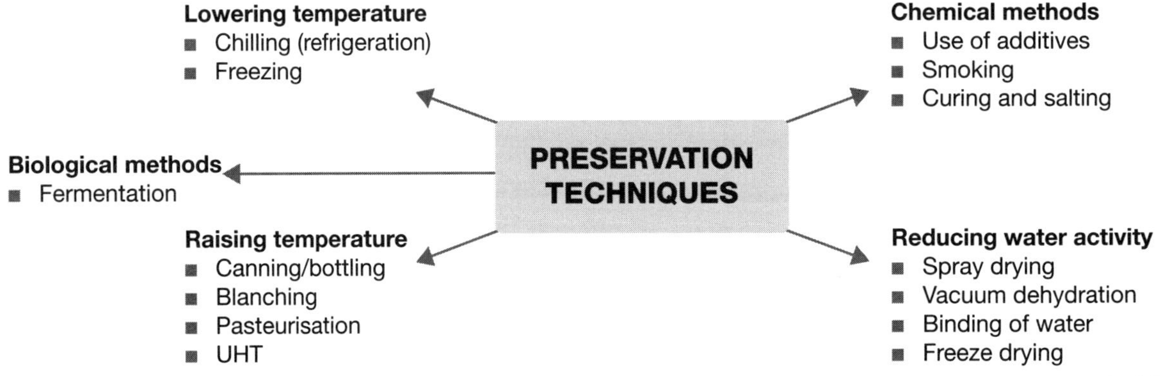

Figure 2.3 Methods of preserving food

Raising of temperature

Canning/bottling

Canning is a heat treatment process that aims to **heat the food** to the point where commercial sterility is achieved yet the food is still edible. Food is sealed into **sterile containers** to prevent micro-organisms entering after treatment of the food. Enzymes are also destroyed in this process.

There are two types of canning processes:

- **Aseptic canning**: foods are first heated and then placed in a sterile container and closed
- **Conventional canning**: foods are placed in the container, sealed and then heated.

In both processes the cans are heated in retorts (large pressure cookers) for the length of time needed to destroy the micro-organisms. Temperatures used are between 105 and 140 °C. The length of time for heating depends on:

- pH of the food—the higher the pH, the more alkaline the food. In this instance the heat treatment is more severe as micro-organisms such as bacteria spores can grow in high pH environments.
- Density of the food—the less dense the can contents, the less the heating time.
- Size of the container used—larger cans require a more intense heat process.

After heat processing, cans are cooled rapidly to avoid the 'critical temperature zone' (4 to 60 °C).

Pasteurisation

Pasteurisation is exposing foods to a high temperature, usually more than 60 °C and less than 95 °C in order to destroy certain micro-organisms. Foods that are preserved by this method include fruit juice and milk.

Temporary	Permanent
Low severity heat treatment	**High severity heat treatment**
■ Food product is heated to 72–73 °C and held for 15 seconds ■ This method is also known as high temperature/short time method (HT/ST) ■ Examples of products include milk	■ Makes food product shelf stable ■ Examples of products include fruit juices

The great **advantage** of pasteurisation is that organisms such as bacteria are destroyed with no significant change in the chemical composition, flavour or texture of the milk.

Ultra Heat Treatment (UHT)

UHT is a process used to preserve milk so that it is shelf stable, that is, will **not require refrigeration** for storage until after opening. Milk is treated to about 140 °C for 3–5 seconds to kill all heat-resistant bacteria. Milk and fruit juices are subjected to this process.

Activity 6 A PAGE 66

Outline the reasons why pasteurised milk requires refrigeration but unrefrigerated UHT milk can be stored on the shelf before opening.

Blanching

This is a process of immersing vegetables in **boiling water** to destroy enzyme activity and pathogenic micro-organisms.

Lowering of temperature

Chilling

Chilling is the storage of foods above freezing point and below 15 °C. This process **slows** the growth and activity of enzymes and micro-organisms. The lower the temperature, the slower the enzyme changes and microbial growth that occurs in food. Foods using this process include fruits, vegetables, meat, fish, milk and cheese.

Freezing

Freezing is the storage of foods between the temperatures of –15 °C and –30 °C. Freezing inactivates but does not destroy enzymes. It also changes water to ice, so it is not available for microbial growth. The **speed of freezing** a food product is critical to the **quality** of the food product. The slower the process of freezing a product, the larger the ice crystals that can form. Upon thawing, cell rupture can occur, thus causing deterioration of the food product.

Reducing water activity

Micro-organisms and enzymes require moisture for growth. The principle behind moisture control is to **remove water content** so it is not available for microbial and enzyme growth.

Spray drying

Spray drying is the removal of water under controlled conditions in a **spray drier**. The product is sprayed and circulated by hot air to remove water. In the classroom the same principle applies for sun-drying or air-drying. Manufacturers aim to dry products at which moisture content is 5–6%, as moulds and bacteria can grow in as little as 13–20%. Milk and eggs are dried to powders using this method.

Vacuum dehydration

Water is removed from liquid foods by the application of **heat in a vacuum**. Evaporated milk and condensed milk are produced using this process.

Freeze drying

The food is rapidly frozen, broken into smaller pieces and placed on shallow trays. Using a vacuum environment and gentle heat, the ice crystals sublimate (change from a solid to a vapour without passing through the liquid phase) which **leaves dehydrated pieces**. Foods processed in this manner include instant coffee, tea, soup and mushrooms.

Binding water

The amount of water present in a food can be reduced by **adding soluble materials** such as sugar and salt to the food:

- **Salt** acts as a water binder by reducing moisture in the solution, and also drawing out water in the bodies of micro-organisms, retarding their growth. Salt is used on vegetables for this process.
- **Sugar** is used as a water binder in jams, condensed milk and syrup.

Chemical methods

Chemical additives

These are preservatives that are added directly to the food product or appear due to processing techniques. If the **concentration of additive** is high enough, the micro-organisms will be killed. Chemical additive examples are sulfur dioxide, benzoic and sorbic acid, antibiotics and antioxidants. Foods that use this process include vegetables, cheese products, ham, bakery goods, fruit juices and jams.

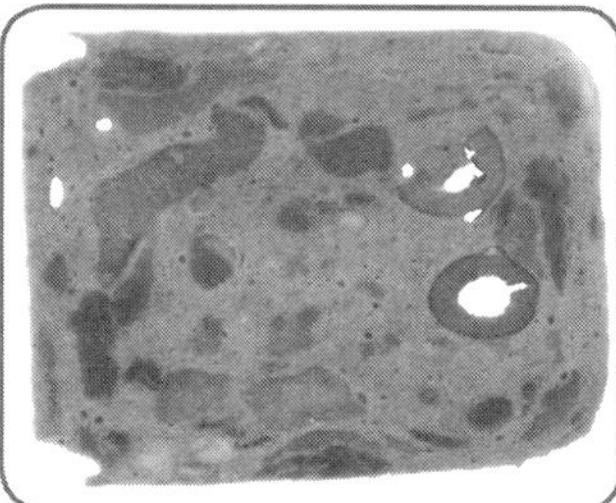

Smoking

The smoking process combines **heat and anti-bacterial agents** to destroy bacteria in meat products. It is also used to inhibit enzyme activity in meats.

Curing and salting

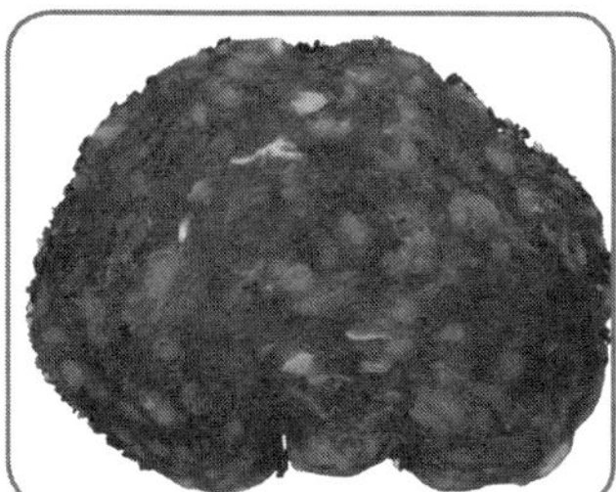

Curing uses the combination of nitrite, nitrates and salt for preservation. The main method of preservation is the action of the **salt-binding moisture**. Deli meats such as salami are preserved in this way.

Biological methods

Fermentation

Fermentation is a preservation process that produces **alcohol or acid** through the action of micro-organisms such as yeast, mould and acid-producing bacteria. In wine making, the build-up of alcohol when fruit is combined with the micro-organism yeast eventually stops the growth of the yeast and stabilises the wine. Other foods that use this process include yoghurt, cheese, soy sauce and bread.

Activity 7 A PAGE 67

Complete the following table by summarising the preservation techniques. The first technique has been done as a guide.

Preservation method	Food examples	Preservation principle: why does this method work?	Additional information
Freezing	■ Meat ■ Fruit	■ Slows enzyme action ■ Makes water unavailable for microbial growth as it is ice	■ Freezing kills up to 80% of microbes ■ Blanching is used in vegetables to reduce enzyme action
Sun drying			
Bottling			
Salting			
Fermentation			
Chilling			

2.3 Packaging, storage and distribution

A manufacturer's choice of packaging depends on a number of factors:

- **Machinery and equipment** available and the additional costs of investing in new machinery for a change of packaging
- **Costs of packaging** including materials, supply, labelling, filling, transport, storage and the package size and stackability and **related cost of supermarket space**
- Handling **properties** for storage and distribution
- **Marketability** of the package.

Functions of packaging

Packaging has five major functions. Packaging is designed to:

- **Contain** the product
- **Protect** the product from:
  - physical/mechanical damage
  - vertical/horizontal storage damage
  - the elements, such as moisture.
- Offer **convenience for the consumer**, in particular:
  - the ability to use the package as a part of food preparation as with microwave meals
  - the ability to recycle the container, that is, use packaging for a secondary purpose.
  - the ability to dispense a limited quantity of the product, such as in tubes.

- **Preserve food** by:
  - preventing contamination by micro-organisms
  - preventing moisture spoilage
  - allowing fresh fruits and vegetables to breathe/respire
  - protecting food from UV rays/light
  - helping prevent rancidity in fats.
- **Inform consumers** and markets about the product by:
  - identifying the product and brand
  - showing quantity
  - explaining features of the product
  - providing directions for use
  - providing information required by law, that is, use-by dates, nutritional information, ingredients list, name and address of the manufacturer.

Materials and packaging design are chosen according to these factors:

- The food must **suit the packaging**.
- The food must be **protected and preserved**.
- The package must be easily **identifiable** to the consumer.
- The food must **not react** with packaging material.
- The package should be **convenient to use**.

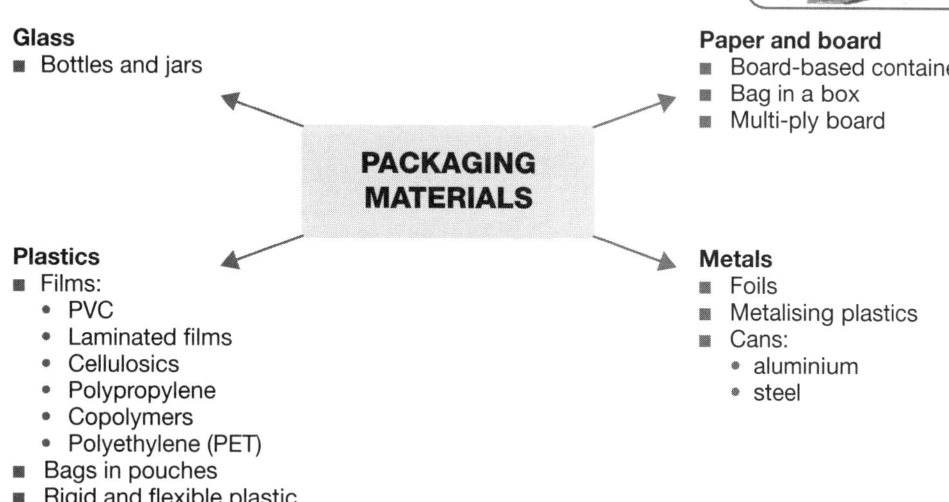

Figure 2.4 Types of packaging materials

Material	Characteristics or advantages	Food application
Metal	■ Steel and aluminium are the materials used ■ Inside, the cans are coated with tin and sometimes plastic or lacquer to prevent foods reacting with the steel. Highly acidic products will require coating of the inside of the can. ■ Cans made from aluminium: ● are cheap ● provide good protection ● prevent moisture loss ● are easy to handle during processing—quick and cost effective ● are displayed and stored easily ● have long storage periods. ■ Steel is virtually unbreakable	■ Steel is used for solid/semi-solid foods ■ Aluminium is used for liquids
Aluminium foils and laminations	■ Most foils are made from aluminium ■ Laminations are a combination of foils and paper or plastic ■ Both are light and flexible ■ Paper or plastic add strength to the foil ■ Foil provides good barrier properties for moisture and oxygen	■ Health bar wrappers ■ Potato chip packets
Glass bottles and jars	■ Glass is inert (i.e. does not react with foods) ■ Odourless and hygienic ■ Can be easily sterilised ■ Is strong ■ Can be reused and recycled ■ A large variety of foods can be packaged this way ■ Contents can be easily seen	■ Liquid, semi-solid and solid foods
Paper and cardboard ■ Board-based containers ■ Bag in a box ■ Multi-ply boards	Paper can be treated to suit the type of food being packaged (e.g. use of waxed papers for foods high in oil) ■ Packaging is cheap ■ Paper/cardboard is inert ■ Paper/cardboard is recyclable ■ Paper/cardboard can be moulded to suit the shape of the food (e.g. egg cartons) ■ Paper/cardboard is sometimes combined with metal for added protection	■ Most foods as mentioned ■ Examples include the food canisters used for Pringles chips
Plastics ■ Laminated ■ Cellulosic ■ Poly-propylene ■ Copolymers ■ PET ■ PVC	■ Rigid packaging is light and strong ■ High impact resistance ■ Available in a variety of colours, shapes and sizes ■ Relatively cheap to produce ■ PVC and PET is commonly used in rigid plastics ■ Flexible plastic packaging is versatile and used for a variety of packaging uses ■ Laminates are the combination of two or more materials that are joined by adhesive or heat ■ Laminates are strong and inert due to the use of an impenetrable plastic (e.g. ethylenevinyl alcohol on the inside)	■ Food pouches ■ Shrink wrap for transport of products ■ Covering of meats ■ Modified atmosphere ■ Packaging of vegetables

Current developments in packaging

The need for new packaging technologies can be linked to an increase in:

- Innovative **technological developments**
- The need for food to be **available all year** round
- Urbanisation, leading to a **decrease in home-grown foods** and greater need for people to access a safe, external supply of fresh, nutritious food
- **Affluence** and participation in paid work outside the home resulting in the need for **convenience** and more efficient packaging
- **Energy costs** and raised awareness of **environmental issues**, leading to the demand for more energy-efficient packaging of foods, and **reduction of over-packaging**
- **Age** of population, leading to demand for the provision of **easy opening packaging systems** with tamper-evident closures
- Demand for **consumer safety** also favours **tamper-evident packaging**
- Legislation on the **labelling requirements** of packages.

New packaging developments are in the areas of:

- Cans
- Active packaging
- MAP packaging
- *Sous vide*
- Plastics development.

Cans

Steel alloys used in can production are being developed so that cans can be produced with **thinner walls**, thus **reducing costs** and allowing for the introduction of the ring-pull can.

Plastic cans have been developed which are:

- Capable of being heated up to 120 °C while still maintaining sound structure
- Designed to provide strength and gas barrier properties
- More expensive than conventional cans, but allow consumers to view the food product (this is often an advantage for gaining consumer acceptance of a new product).

Active packaging

Active packaging refers to the use of materials designed to **interact with the air in the package**, so creating an atmosphere that results in **longevity** of the food product.

Active packaging involves the use of sachets or films to either remove or add gases to the **package headspace**. The principle behind this is that fresh foods, when packaged, are still biologically active and produce moisture due to humidity. Consequently, oxygen levels decrease and carbon dioxide increases. This environment causes rapid food spoilage.

In active packaging, **scavengers** are materials used that **remove gases from the package**:

- O_2 scavengers **reduce oxygen levels** to delay surface browning of foods, such as dried apples, to delay rancidity in foods containing vegetable oils, such as potato chips, to retain colour in packaged cured meats and to prevent mould in packaged cheese.
- Ethylene scavengers are used to **trap ethylene** produced by ripening of fruits and vegetables which delays the ripening process. They are especially effective when packaging fresh fruit and vegetables for export.
- Water vapour absorbents are used to help control the presence of **condensation** in fruit and vegetables, which encourages mould growth.
- CO_2 scavengers are used to **absorb carbon dioxide** and so reduce gas build-up in packaging. Often used in coffee packaging.
- CO_2 and O_2 scavengers are often combined together.

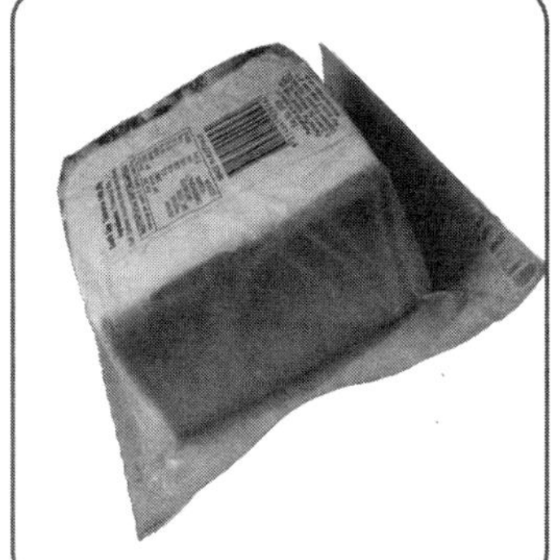

MAP packaging

MAP packaging works on the principle that the **starting atmosphere** in the package is adjusted to produce a gas mix that will increase shelf-life. There are two methods used for this process, depending on the food to be packaged:

- The headspace of the package is filled with **gas** or a mixture of gases that will extend shelf-life of the food. The packaging material used will not allow the gases to escape; for example, a combination of CO_2 and nitrogen is used for fresh-baked products.
- A second form of MAP packaging uses a **film** that allows movement of gases and water into and out of the package. An example of this is the use of packaging that allows O_2 in and CO_2 out. The ethylene (a plant hormone that stimulates ripening and ageing) produced in this respiration process is absorbed by the film, rather than building up in the package. This process is used extensively with fruits and vegetables as it slows the ripening process.

Sous vide—vacuum cooking/packaging technology

A cooking technology based on **minimal heat treatment** of food, *sous vide* is used to extend the shelf-life and keeping qualities of fresh food. Foods are cooked in sealed, air-evacuated heat stable pouches or thermoformed trays so that the natural flavour, aroma and nutrient quality are retained by the product. After the cooking procedure the food product is **blast chilled** to 3 °C. Foods prepared in this way are often packaged with the use of polypropylene and polyethylene laminates and will have a shelf-life of up to six weeks.

The **advantages** are that this process:

- Enhances the **natural taste** of food
- Accentuates the **tenderness** of meat
- Requires **no cooking in fat**
- Retains **maximum nutrient** content
- Provides **easily prepared meals**; for example, reheating for 10–15 minutes in boiling water or 4–5 minutes in the microwave.

The **disadvantages** are that this process involves:

- High risk of consumer **temperature mistreatment**, that is, not heating the food product sufficiently or overheating it
- Consequent **microbial hazards** if reheating is not done sufficiently.

Plastics/polystyrene developments

Dual ovenable plastic is capable of withstanding both microwave heating as well as conventional oven heating. It is expensive but provides convenience. Instead of ozone-depleting agents being used in the production of foam trays, a new process has been devised. **CO_2**, which is used as the foaming agent, has **no destructive effects on the ozone layer**.

Activity 8 A PAGE 67

(a) Identify a food product that uses each of the packaging techniques just mentioned.

(b) Choose a food product that is packaged in a variety of ways, such as orange juice. Analyse each packaging material as to its acceptability for that particular product. Write up your analysis as summary notes.

Storage and distribution during food manufacture

Storage and distribution need to be considered separately.

Storage

Storage in food manufacture refers to:

- Keeping of **raw materials** before processing
- A **delay stage** during the processing method where the product has to be stored for a process to occur, such as bread dough rising
- When the final product is stored **before distribution**.

Conditions for storage are dependent on the nature of the product being stored:

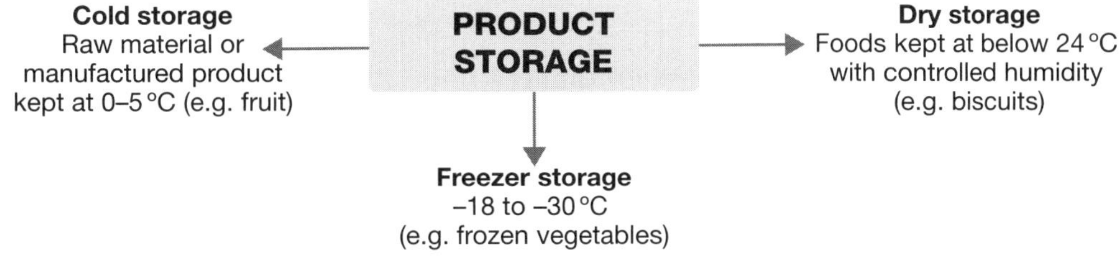

Figure 2.5 Product storage

Distribution

The distribution of a product is the movement of goods from one site to another.

For goods to be distributed successfully, these aspects of the product need to be considered:

- The type of material being moved; for example, if perishable it would require refrigerated transport
- The packaging used so that the product is delivered in quality condition
- Cost-effective movement of the goods that will benefit manufacturer and consumer.

Distribution needs to be considered from two angles, the distribution channels and the distribution systems.

Distribution channels refers to where the product goes before it gets to its final destination.

Distribution channels

Distribution channels can be modelled as:

The distribution channel used is dependent on the **nature** of the product and the **size** of the company. Effective distribution channels result in **less waste** due to better management of:

- Labour
- Handling
- Storage space
- Time.

A typical distribution system could be:

Raw materials transported and stored as needed

↓

Food product processed and packaged (e.g. MasterFoods)

↓

Product delivered to warehouse (e.g. Davids Holdings).
Warehouse receives bulk supplies from food manufacturer and divides supply into smaller units. Distribute to retailers (e.g. Coles).

Chapter summary

The three areas of content that students should know are:

1. **Production and processing of food**

- Quality and quantity control in the selection of raw materials for food processing
- Role of food additives in the manufacturing process
- Characteristics of equipment used in different types of production and the factors influencing its selection
- Production systems used in the manufacture of food, such as small-scale, large-scale, manual, automated and computerised
- Quality management considerations in industrial practices to achieve safe foods for public consumption, such as hazard analysis and critical control point (HACCP), occupational health and safety, and hygiene.

2. **Preservation**

- Reasons for preserving foods, such as safety, acceptability, nutritive value, availability and economic viability
- Causes of food deterioration and spoilage:
  - environmental factors (infestation, oxygen, light and water)
  - enzymatic activity
  - microbial contamination (mould, yeast and bacteria)
- Principles behind food preservation techniques, such as temperature control, restriction of moisture, and use of chemicals
- Preservation processes, including canning, drying, pasteurising, freezing and fermenting.

3. **Packaging, storage and distribution**

- Functions of packaging and types of materials available
- Current developments in packaging, such as active packaging, modified atmosphere packaging and *sous vide*
- Legislative requirements for packaging and labelling
- Storage conditions and distribution systems at various stages of food manufacture.

Revision questions

Multiple choice

A PAGE 68

1. Raw material specifications are most essential for
   A the quality control of raw materials.
   B the correct processing procedures.
   C all products to be produced the same.
   D reducing waste.

2. Humectants are used in food processing to
   A inhibit growth of micro-organisms.
   B prevent oxidation.
   C retain the nutritive value of the food.
   D absorb moisture from the atmosphere so the processed food does not dry out.

3. Food manufacturers use HACCP to
   A identify potential hazards within production of a food product and solutions to deal with them.
   B find insects in raw materials.
   C identify poor food produced.
   D keep accurate records about specific food production.

4. Food deterioration and spoilage is caused by
   A micro-organisms.
   B enzyme activity.
   C environmental factors.
   D all of the above.

5. An example of raised temperature preservation is
   A fermentation.
   B pasteurisation.
   C curing.
   D smoking.

6. The nature of large-scale food production mainly involves the use of
   A computerisation.
   B manual labour.
   C HACCP.
   D production runs.

7. The following ingredient list appears on a food label:

   Chicken, Wheat Flour, Canola Oil, Breadcrumbs, Protein Concentrate, Rice Flour, Dehydrated Potato, Mineral Salt (451), Gluten, Sugar, Emulsifiers (471,481), Preservative (282), Water added.

   From the ingredient list, it is apparent that the food product
   A should be stored at room temperature.
   B should be stored in a freezer.
   C is a solid, dehydrated product.
   D is a free-flowing, dehydrated product.

8. The main preservative principle in pasteurisation is the
   A exclusion of air.
   B reduction of moisture availability.
   C use of high temperatures.
   D addition of acid to change pH levels.

9. A preservation technique that reduces water activity is
   A chilling.
   B spray drying.
   C smoking.
   D fermentation.

10. If a packaging material is inert, it does not
    A break.
    B react with the packaged food.
    C preserve the food.
    D allow the consumer to see the product.

11. A specific process that transforms raw materials into manufactured products is
    A separation.
    B freezing.
    C mixing.
    D all of the above.

12. The responsibility of the employer under occupational health and safety guidelines is to
    A provide minimum training for employees.
    B provide meals for employees.
    C maintain processing equipment only when broken.
    D instruct or train about safety issues.

13. Food preservation can best be defined as
    A the processing of food to eliminate the conditions that cause spoilage.
    B changing food to improve flavour.
    C reducing the shelf life of food.
    D cooking food in hygienic conditions.

14. A function of packaging is to provide convenience for the consumer. An example of this function would be
    A the ability to recycle the container.
    B to protect from physical damage.
    C to provide information to consumers.
    D to protect food from UV light.

Short-answer responses

A PAGES 68–69

15. Name a preserved food and describe the packaging used. (1 mark)

16. (a) Identify four functions of packaging. (4 marks)
    (b) Describe how this type of packaging preserves the food. (4 marks)

Structured extended response A PAGE 69

17. (a) Identify a raw food material and discuss the preservation techniques that could be used to preserve this food. (8 marks)

    (b) Explain the preservation principles behind each process outlined. (7 marks)

Answers Food **Manufacture**

Activity 1

Term	Meaning
Active packaging	The use of packaging materials designed to provide a barrier to external contaminants and also to interact with the air in the package, creating an atmosphere that results in longevity of the food product
Automated	Use of machines to handle and control processing from raw materials to the finished product
Computerised	Use of sensors in production to enable large amounts of data to be stored to ensure production specifications are met
Critical control points	Parts of the production process identified to be a point where a hazard could occur
Enzymatic activity	Chemical substances that act as catalysts in chemical reactions in food, such as browning of meat and ripening of fruit and vegetables
Equipment	A collection of tools used for processing
Food additive	A substance that is added to food to enhance the food product and is not normally consumed on its own
Food deterioration	The breakdown of food so that it is unfit for consumption
Food manufacture	Activities that use manual, automated and computerised processes to produce a food product from a raw material
Food preservation	The processing of food to eliminate the conditions that cause food spoilage and deterioration
HACCP	A quality management system that identifies potential hazards within the production of a specific food product
Integral processes	Processes that are a part of the whole production process
MAP packaging	The starting atmosphere in the package is adjusted to produce a gas mix that will increase shelf life
Microbial contamination	Food spoilage caused by bacteria, yeast and moulds
Preservation processes	Processes used to prolong the optimum quality of food
Principles of food preservation	The control of factors that contribute to food deterioration
Processed food	A food product that has been produced from raw materials
Production run	Conversion of raw materials into a final product within a certain timeframe
Quality assurance	Guarantee that standards in food manufacture are maintained so that food quality is not compromised
Quality control	Methods used to ensure that product quality is maintained
Raw material	Any food material that is used in the production of another processed food product
Sous vide	Food is cooked, vacuum packed and chilled for later use
Specification	A detailed description of materials used

Activity 2

Thickener: Used to make the food product thicker. In this product it would improve the quantity of the product.

Mineral salts: Added to improve the texture and mouthfeel of the product. In this product the mineral salts would help to make the product smoother.

Food acid: Used to maintain a constant level of acidity and to produce a sharp taste. Can enhance the flavour of the product so it does not taste bland.

Vegetable gum: Used to impart consistency and texture. This would work with thickeners and mineral salts.

Colour: Used to improve the appearance of the product and encourage consumer acceptance of it.

Activity 3

Manufacturing process for ice-cream.

Industrial	Domestic
■ Delivery of raw materials and inspection. ■ Milk and cream products blended and homogenised so the milk components will not separate. ■ Mixture is then pasteurised—milk product held at a high temperature for a period of time to destroy bacteria. ■ Mix is whipped and frozen. Air is pumped into the mixture while the whipping is being completed. Quick freezing of the mixture helps to prevent formation of large ice crystals. ■ Other ingredients such as fruit or nuts are added at this stage. ■ Ice-cream is packaged and hard frozen. ■ Freezer storage and distribution.	■ Dissolve gelatine. ■ Beat milk ingredients with beaters until well blended. Mix in gelatine. ■ Freeze in freeezer trays for four hours. ■ Take out, break up and beat with electric mixer until double in size. ■ Refreeze.

Activity 4

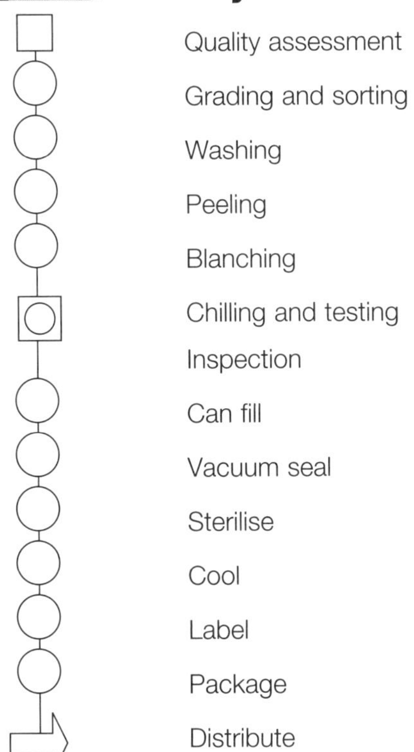

Activity 5

Food production company is Sanitarium. QC is quality control procedure.*

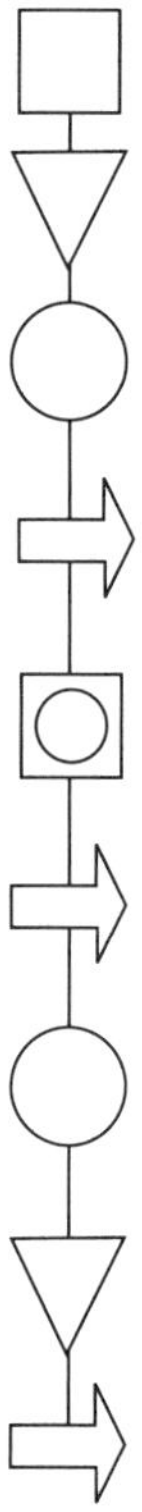

Wheat is inspected. **QC of raw material through scanners.*

Wheat is stored below 20 °C. **QC of checking storage temperature.*

Wheat's physical and chemical characteristics are changed due to heating.

Wheat travels in a steady measured stream to the mills.

Physical characteristics changed at this stage (moulded to form shapes) and inspected. **QC is inspection of product. Would be tested for hardness, size etc.*

Biscuits are then dropped onto another conveyor belt.

Physical characteristics changed (colour) as shapes are toasted. The cereal is then packaged. **QC of colour checked and temperature used for toasting.*

Cereal is stored.

Transported to the wholesaler for distribution.

Activity 6

(a) Higher use of temperature for UHT ensures destruction of bacteria.

(b) Aseptic packaging is used with UHT milk. This involves the package and the food product being sterilised separately, and then filled in sterile containers. This process ensures less risk of contamination and allows the product to be stored without refrigeration.

(c) Pasteurised milk requires refrigeration as it is not aseptically packaged.

Activity 7

Preservation method	Food examples	Why does this method work	Additional information
■ Freezing	■ Meat ■ Fruit	■ Slows enzyme action ■ Makes water unavailable for microbial growth (as it is ice)	■ Freezing kills up to 80% of microbes ■ Blanching is used to reduce enzyme action
■ Sun-drying	■ Fruit	■ Removes moisture so it is not available for microbial growth	■ Moulds and bacteria can grow in 13–20% moisture content. Drying reduces levels to below this percentage
■ Bottling	■ Fruit ■ Vegetables ■ Juice	■ High temperatures destroy microbes and enzymes ■ Seal prevents entry of microbes	
■ Salting	■ Fish ■ Meat	■ Salt makes water unavailable for microbes	■ Least used method compared to others
■ Fermentation	■ Cheese ■ Yoghurt ■ Pickles	■ This involves the use of acid or alcohol to preserve foods	■ Mould and yeast are organisms used in the fermentation process
■ Chilling	■ Milk ■ Fruit ■ Meat ■ Vegetables	■ Slows enzyme action	■ 0–4 °C is chilling temperature

Activity 8

(a)
- Cans (e.g. tomatoes)
- Active packaging (e.g. cheese)
- MAP packaging (e.g. salad mix)
- *Sous vide* (e.g. prepared beef stroganoff)

(b) **Orange juice**

Glass—recyclable and reusable material
- Allows consumer to see product
- Vitamin C content could be harmed by light
- Inert
- Breakable
- Heavy.

Foil laminate board—recyclable
- Easily stored
- Laminate makes foil inert. Foil increases shelf-life as oxygen is less likely to penetrate, and also strengthens the package.
- Lightweight
- Quickly chilled.

PET plastic—recyclable and reusable
- Allows consumer to see product
- Inert
- Resistant to breakage
- Light could affect vitamin C content.

Multiple choice

1. **C** Specifications are put in place by food manufacturers to ensure production continuity of a quality product.
2. **D** The function of humectants is to absorb moisture.
3. **A** HACCP is a recognised quality control system in food production.
4. **D** Micro-organisms include bacteria, yeast, moulds and viruses, environmental factors include infestation, oxygen, light and water; enzymes are chemical substances present in foods and are responsible for ripening and maturation of foods.
5. **B** Pasteurisation involves the use of high temperatures (between 60 and 95 °C) to reduce microbial activity.
6. **A**
7. **B** The product should be stored in the freezer as it has chicken as one of its main components. All other answers do not reflect the inclusion of a fresh ingredient.
8. **C** Pasteurisation is the process of exposing milk to a high temperature in order to destroy micro-organisms.
9. **C** Water is one of the environmental factors required for microbial growth. Spray drying reduces water content so removing this factor.
10. **B** An inert material is one without active properties so it does not react with food. Glass is an example of this.
11. **D** All processes listed can be used to transform raw materials.
12. **D** It is the responsibility of the employer to ensure employees are trained in all aspects of occupational health and safety for their specific workplace.
13. **A** Food preservation is the processing of food to eliminate the conditions that cause food spoilage.
14. **A** Recycling provides convenience for the consumer as it reduces waste and unnecessary build-up of waste.

Short-answer responses

15. Ardmona pears. Steel can is the packaging.
    1 mark
    Named preserved food and described package

16. (a) (Candidates choose four out of the following five answers)
    1. Contain the product
    2. Protect the product from
        - Physical damage
        - Vertical/horizontal damage
        - Elements (e.g. moisture).
    3. Convenience
    4. Preserves food
    5. Informs consumers and markets the product.

    4 marks
    1 mark per function

(b) (Candidates choose four out of the following five answers)

1. Packaging prevents contamination by micro-organisms (bacteria, yeasts and moulds), moisture spoilage and oxidation.
2. The use of steel ensures an effective protection barrier as steel is strong and can be made tamper resistant.
3. Welded seams are used to increase strength of the can and to protect contents from contamination due to broken joins.
4. Steel cans can be heated to extremely high temperatures, between 105 and 140 °C, so ensuring sterilisation of cans and contents. The use of such temperatures also ensures the destruction of micro-organisms.
5. Cans are sealed to prevent contamination of contents.
6. Cans can be stacked during storage, thus preventing damage due to poor storage and handling procedures.

4 marks
Two major features
1 mark for description of feature
1 mark for explanation of how the feature assists in the preservation of food

Structured extended response

17. (a) *Identify a raw food material (e.g. tomatoes)* *1 mark*

Preservation techniques *7 marks*

Canning – Canning is a heat treatment process that aims to heat the food to the point where commercial sterility is achieved. In aseptic and conventional canning the cans are heated in retorts to temperatures between 105 and 140 °C to destroy micro-organisms. Cans are then cooled rapidly to avoid 'critical temperature zone' 4–60 °C.

Chilling – Chilling is the storage of food above freezing point and below 15 °C.

(b) *Preservation principles* *7 marks*

Canning – Canning uses the principle of temperature control to destroy micro-organisms such as bacteria and mould. This technique uses heating of food to between 105 and 140 °C. This preserves the food as microorganisms cannot survive such high temperatures.

Chilling – The temperature zone between zero and 15 °C slows the growth and activity of enzymes and micro-organisms. The lower the temperature the slower the enzyme changes hence the tomatoes ripening process is slowed down and over-ripening does not occur.

3 – Food Product Development

Food product development is an integrated system involving expertise in the fields of marketing and manufacture. The food product development process applies knowledge and skills developed through studying of a range of areas, including nutrition, food properties and food manufacture.

Outcomes

On completing this chapter a student should be able to:

1. Justify processes of food product development and manufacture in terms of market, technological and environmental considerations (H 1.3)
2. Develop, prepare and present food using product development processes (H 4.1).

Activity 1 A PAGE 90

Following is a list of terms you should know as a result of completing this module.

Term	Meaning
Decline	
Design brief	
Ecological	
Feasibility	
Line extensions	
Macro environment	
Market research	
Market share	
Maturity	
Me-toos	
Micro environment	
New to the world	
Packaging	
Primary	
Profitable	
Promotion	
Prototype	
Qualitative	
Quantitative	
Screening	
Secondary	
Sensory evaluation	
Specifications	
SWOT	
Target market	

The HSC Examination

In the HSC Examination students could expect to answer questions on Food Product Development to the value of approximately 25% of the paper. Questions may require students to integrate knowledge, understanding and skills developed through studying the entire course. It is therefore difficult to predict the weighting of marks in each section of the paper.

The paper may include:

- Multiple-choice questions
- Short-answer questions (which may include parts)
- A structured extended-response question, which includes two or three parts with one part worth at least 8 marks

OR

- An extended-response question (approximately 600 words).

3.1 Factors which impact on food product development

Factors which impact on food product development can be divided into **external** and **internal** factors.

External	Internal
Economic environment ■ Exchange rates ■ Inflation/recession ■ Unemployment ■ Taxation ■ Salary ■ Importance of the food industry to the Australian economy ■ The economic cycle	**Personnel expertise** ■ Skills levels ■ Transferable skills ■ Changing employment ■ Importance of education and training ■ Flexibility ■ Part-time, casual work
Political environment ■ Government influences/controls (policy and legislation): ● price ● the environment ● education ● working conditions ■ Community influences ■ Lobby groups	**Production facilities** ■ Importance of level of production facilities: ● small volume/one-off ● batch production ● mass production ■ Storage ■ Distribution
Ecological environment ■ Natural resource use ■ Environmentally friendly decisions ■ arming procedures ■ Waste management ■ Packaging ■ 'Clean green' image ■ Consequences of disregard for environmental issues	**Financial position** ■ Stability of the company ■ Balance between expenditure and profit ■ Investment ■ Resource management ■ Product quality ■ Market share/competition
Technological environment ■ Improved productivity ■ Increased choice ■ Relationship between level of production and technology used ■ Cost of technology/cost effective for large-scale production	**Company image** ■ Influenced by: ● reputation ● quality ● marketing strategies ● packaging ● price

External factors (the macro environment)

There are a number of influences on food product development which are beyond the control of **industry**. These are the **macro environment** or **external** factors.

The economic environment

Economic activity usually **fluctuates**. The influences on the fluctuations are complex and several factors determine economic **growth** and **stability**.

Exchange rates

As the value of the Australian dollar fluctuates, there are variations in the appeal of our produce in the global marketplace. When our dollar is strong, primary produce and manufactured goods may not be competitive; however, when the value of our dollar falls there is a huge demand for our foods.

Fluctuating interest rates

Fluctuations in rates create uncertainty for **investors**. Low interest rates may encourage food organisations to invest in **capital**, such as equipment, expansion, improved technology.

Inflation/recession

Inflation/recession involves an increase or decrease in the cost of living. This in turn influences people's **purchasing power**. Fluctuations in available income will influence the type of products purchased. When inflation is high, more **value-added foods** are consumed and more meals eaten away from home, whereas during a **recessionary** period, consumption of **generic brands** or **staple** food items increases.

Unemployment

Unemployment creates an increase in the number of people living on **social security**. These people will have less choice and their food purchases may be limited to less processed products, foods that are in season and less expensive cuts of meat. High unemployment rates are an additional burden on governments as more money is devoted to welfare payments, restricting other government initiatives such as research, improved health care, law enforcement or education.

Taxation

Taxation includes the type of taxes and level of taxation. This impacts on price, available income and purchasing power. The **Goods and Services Tax** (GST) aims to ensure that equal tax is paid by all consumers on the items or services they purchase (user pays).

Salary negotiations and award wages

Income impacts on **spending power**, and generally the higher the salary, the more choice in food products.

The Australian food industry

The food industry is very important to the Australian economy. Some features include:

- 17% of the manufacturing workforce is in the food industry.
- Approximately 180 000 people are employed in the food industry.
- Employment in the food industry is increasing as value-added foods become more important in the marketplace.
- Increased technology and computerisation in the food industry is having a levelling effect on rising employment.

- Foreign ownership and control dominates the food industry.
- Australian families spend about 20% of their income on food.
- Despite Australia's ability to produce sufficient food for its people, we still import a significant quantity of food (A$9 billion in 2007–08).
- Australia needs to be more competitive in the international marketplace to increase the volume of foods exported.

The economic cycle

Economic stability is important to the industry within the Australian economy and in the global marketplace. Fluctuations in the economy are often described as the economic cycle. There are four main stages:

1. A **recession** is characterised by high unemployment, interest rates and inflation and results in frequent instances of business failure.
2. During **expansion**, confidence increases, employment opportunities grow and interest rates and inflation fall.
3. A **boom** is characterised by low unemployment, low interest rates and increased spending.
4. Economic **contraction** occurs as confidence declines with increasing interest rates and spending slows—a recession appears inevitable.

These stages are often **global** situations and therefore may be very difficult for governments to manage by monetary decisions (e.g. raising or lowering **interest rates**, **wage freezes**).

Economic conditions are influenced by the **world economy**. While the **Treasury** and **the Reserve Bank** in particular attempt to **manipulate** fluctuations in the economy, at times it is very difficult to control because of circumstances within the economies of our trading partners.

Activity 2 A PAGE 90

(a) Give examples of how you may change your spending on food items if faced with a recession.

(b) How may an expansion phase influence food related purchasing decisions?

(c) What changes might you make to your food selections if you were living in a boom period?

The political environment

In the policy and legislation section of the Australian food industry chapter we saw how Federal, State and local government can influence food operations. **Controls** in the food industry help to ensure a safe and reliable supply of food. These controls have been instrumental in identifying Australian foods as '**clean green**' products that are highly valued throughout the world.

Government

The political 'systems' influence the following:

- **Price**—levels of taxation, competition, trading rules, licensing, subsidies and tariffs
- **The environment**—air, water, noise, use of chemicals in food production and food additives
- **Education**—food and nutrition education, food standards, and food handling and safety
- **Working conditions**—awards, trading hours, occupational health and safety.

Community influences and lobby groups

The community is able to **lobby** politicians to bring about changes in the industry that are in the interest of the community. At times this pressure comes from organised groups with **vested interest** in changes (e.g. Dairy Farmers) or the pressure may result from **widespread concern** about an issue such as the use of landfill for waste disposal or genetic engineering in food production.

Ecological environment

The use of **natural resources and energy** are important in the **promotion** of a highly valued 'clean green' food supply, particularly in areas of the world unable to match the standards achieved in Australia.

With increased concern for **environmentally friendly** behaviour, food producers and manufacturers are continually checked for environmentally responsible decisions rather than decisions that are purely motivated by profit. Issues include:

- Use of natural resources
- Farming procedures
- Waste management and packaging.

At times, the environmentally friendly methods used are not always the most cost effective but may enjoy **popular support** because of the long-term benefits to the environment. Disregard of **ecological issues** not only results in the scorn of consumers—**political consequences** also frequently apply.

Technological environment

Improved **technology** has led to improved **productivity** and increased **choice**. It has also resulted in enormous change in relation to 'how' food is produced. The size of the operation is significant in determining the level of technology employed. Although technology is expensive, when large-volume production is required technology often becomes **cost effective**.

Internal factors (the micro environment)

There are a number of influences on food product development which are **industry** or **market based**. These are the **micro environment** or **internal** factors.

Personnel expertise

The workforce is becoming increasingly skilled. **Employees** change their employment more frequently than has occurred in the past and the skills and knowledge that people learn in one work environment are often transferable to another. **Generic skills** such as the ability to work in a team, good communication skills, problem solving, computer literacy, demonstrating initiative and so on can all be adapted to a range of settings.

Education and training are key ingredients for a skilled workforce. Many **employers** provide comprehensive **on-the-job training programs** as well as supporting and encouraging staff to undertake **ongoing training** through a range of institutions.

Flexibility is important in most work situations, whether this involves knowing how to perform a number of tasks or willingness to adjust **work schedules** with demand. Opportunities for employees with young families are increasingly available, such as part-time work, job sharing, working from home and so on.

Within the food industry, **expertise** is diverse from production, packaging, transportation, marketing, quality assurance, finance, sales, maintenance, management and so on. Whilst expertise in some sectors is highly **specialised**, **broad knowledge/understanding** can influence trends and decisions and ultimately impact on other aspects of the industry.

Production facilities

The need for production **facilities** will depend on the food product and the **volume of production** (see Australian Food Industry—Level of Operation, p. 9). Some factors include:

- **Small volume and one-off operations**—have limited production facilities and domestic resources can be utilised.
- **Batch production**—includes larger runs of a particular product; sometimes variations in production occur as a result of relatively minor changes in the facilities used, such as flavour changes, single-serve packaging, frozen versus fresh sauces.
- **Mass production**—uses automation and computerisation to ensure production volumes meet a large target market. Equipment used is often single purpose and in continual operation, such as in a Coca-Cola factory.

At all levels of production and sales, **storage costs** are an expense that the food industry must control carefully. To avoid the need for large storage facilities, the most efficient process is **'just in time'** deliveries and movement of stock. This practice requires the availability of resources to meet production with limited stock on hand.

Financial position

The success of an operation is usually judged by how well it manages expenditure and the **margin of profit**.

Influences on the **financial success** of a business include:

- Investment capital
- Resource management
- The quality of the product
- The market share—competition.

Company image

Corporate **image** is influenced by a number of factors, including:

- Reputation
- Quality
- Marketing strategies
- Packaging
- Price.

Recent incidents of industrial sabotage have been significant in their effect on **company images**. Some companies align themselves with '**politically correct**' philosophies that improve their image (environment, health, 'clean green') to potentially **enhance sales**.

Activity 3

Select a food company to research on the internet.

(a) Explain the company image.

(b) Is there a mission statement?

(c) What are their environmental policies?

(d) What are their production rates and facilities?

(e) Conduct a SWOT analysis of the company identifying apparent strengths/weaknesses/opportunities and threats.

3.2 Reasons for and types of food product developments

Health issues

In Australia there are a variety of **diet-related health problems** that cause concern for an enormous number of people. Food products developed to address these disorders will have a significant audience (target market).

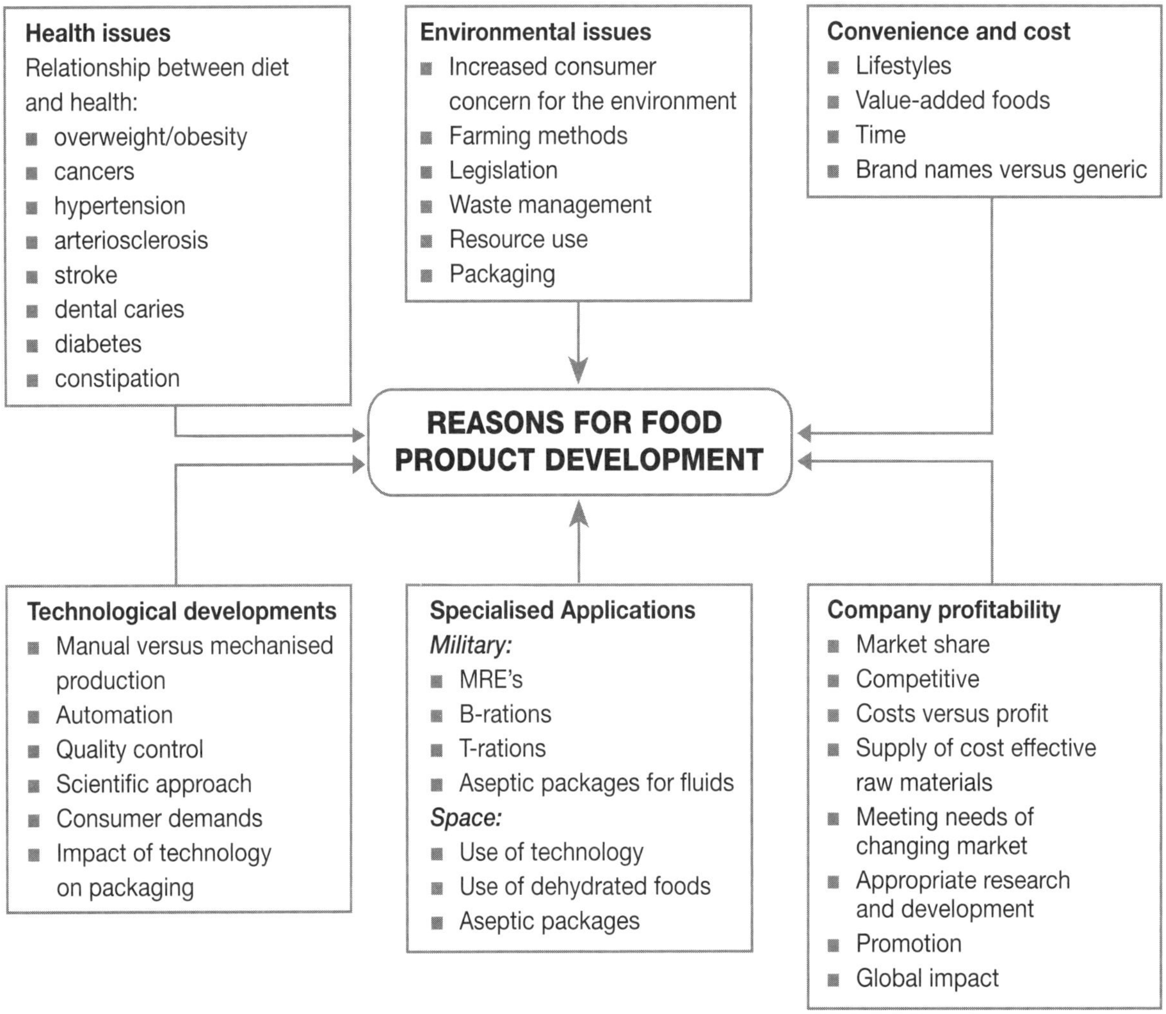

Health issues
Relationship between diet and health:

- overweight/obesity
- cancers
- hypertension
- arteriosclerosis
- stroke
- dental caries
- diabetes
- constipation

Environmental issues

- Increased consumer concern for the environment
- Farming methods
- Legislation
- Waste management
- Resource use
- Packaging

Convenience and cost

- Lifestyles
- Value-added foods
- Time
- Brand names versus generic

REASONS FOR FOOD PRODUCT DEVELOPMENT

Technological developments

- Manual versus mechanised production
- Automation
- Quality control
- Scientific approach
- Consumer demands
- Impact of technology on packaging

Specialised Applications
Military:

- MRE's
- B-rations
- T-rations
- Aseptic packages for fluids

Space:

- Use of technology
- Use of dehydrated foods
- Aseptic packages

Company profitability

- Market share
- Competitive
- Costs versus profit
- Supply of cost effective raw materials
- Meeting needs of changing market
- Appropriate research and development
- Promotion
- Global impact

Figure 3.1 Reasons for food product development

Activity 4 A PAGE 91

Complete the table by identifying the health issue linked to the following product lines and listing examples of food and eating strategies to deal with the issue.

Product line	Health issue	Food and eating strategies
High-fibre	■ Bowel cancer	■ Wholegrain breads and cereals ■ Fresh fruits and vegetables ■ Primary rather than processed foods
Low-fat		
Low-salt		
Low-sugar		
Cholesterol-free		
Nutrient-enriched		

Dietary considerations

Australian society has a large number of diet-related health issues which impact on the types of food products developed. Within the marketplace consumers are looking for foods that will prevent or protect them from the range of illnesses that have developed as a result of ongoing poor food choices. Consequently a wide range of foods have been developed that target the cause of many diet-related diseases. These include reduced fat and sugar; fibre enriched; fortified or enhanced foods and foods which have been modified in some way to alter the original composition.

An example of a primary food that has undergone this level of modification is milk:

- Fortified or enhanced milk (calcium and or vitamin D added)
- Reduced-fat milk (low-fat milk has less than 1.5% fat; skim milk has 0.15% fat)
- Modified milk (Omega-3; dietary fibre; plant sterols).

In addition, non-dairy 'milks' are promoted as a 'healthy' substitute for cow's milk. Made from nuts, legumes or grains (e.g. soy), they are free of cholesterol. Calcium, vitamin D and iron are often added to fortify the beverage to make it comparable to cow's milk.

Activity 5 A PAGE 91

Select and describe another food that has been developed to address dietary considerations.

The environment

Many consumers are concerned about environmental issues and are making choices in their lives to have a positive impact on the environment. This has changed methods of farming with a trend away from the use of pesticides and fertilisers that upset the chemical balance. **Organic farming** has become very popular, as has the use of natural means to enhance yields and crop production.

Food manufacturers are required to comply with **legislation** in relation to clean air, water and noise pollution. Wise energy use and recycling waste materials are not only **cost effective** but also **environmentally responsible**. Thoughtful use of resources not only protects the environment, it also sustains the resources' existence for **future generations**.

The recommendation of **reduce, reuse and recycle** has impacted on industry. Less packaging materials are being used, packages have more than one use and much more effective recycling programs are in place.

Consumer demands such as convenience foods and cost

Changing lifestyles have led to increasing demands for **quick**, **easy** and economical solutions to food decisions.

The demand for food that is quickly and easily prepared has increased as has the **percentage of money** spent to meet this need. The demand for more **convenience** has also led to more flexible shopping hours and an increase in self-service (the development of supermarket-style shopping).

Value-adding increases costs but often reduces the time taken in food preparation. As cost increases, some consumers will be forced to make **economic decisions** about the food products they consume. Name brands and **generic** products compete in the marketplace. At times a consumer will favour a **name brand** but economically find a generic alternative very attractive. Other consumers who are very brand loyal are prepared to pay a premium price for what they believe to be a superior product.

Societal changes including increasing ageing population, single-person households and longer working hours

As women join the paid workforce in increasing numbers, **men** are taking a more **active role** in shopping for, and preparation of, food.

Meals eaten at home have become more varied with a renewed interest in cooking/recipes. This interest has been reflected in the growth of recipe books and cooking equipment in the marketplace. Television programs that focus on **home cooking** have been an integral part of this trend. **Home meal replacement solutions** provide an alternative to meals prepared from scratch and an option that involves less preparation time and washing up.

As Australians are living longer there are more **aged** people in the community. These people are a powerful target market in the development of food products. Digestion issues may impact on preferences for less spicy foods as well as foods that are more easily chewed and swallowed.

Frequently, aged people are living alone and add to the demands of **single-person households** (people marrying later or resulting from the high incidence of marriage breakdown). This group often demands food that can be purchased in small portions and is easily or fully prepared. **Single-serve** and **heat and eat** options have developed to cater for single households that may be time or skill poor.

The changing world of **work** has also resulted in less time at home and therefore less time available for food purchases and preparation. Many more **meals are eaten away from home** (including breakfast). Tastes have become more sophisticated and these demands have been met with an increase in the variety of foods provided in a diverse range of eating establishments (takeaway and restaurants). The cost of these meals is as varied as the selection of foods and is responsive to the disposable income allocated to this market.

Activity 6 **A PAGE** 92

List lifestyle changes that have led to the demand for convenience and value for money in food items.

Technological developments

Manual labour is increasingly being replaced by **machinery** that is quicker, more reliable, less demanding and more cost effective. The technology is often **automated**, giving **consistent** quality and is often computerised, providing **quality control checks** as part of the production process.

All aspects of the food industry have been altered as a result of a more **scientific** approach to food production. Foods are enhanced by the use of **food additives**, **genetic engineering** is im-proving qualities of raw materials and **functional foods** are being developed to address specific dietary needs. These all serve to provide a food supply that is responding to consumer demand rather than simply the provision of food items from which consumers must make choices.

Advances in technology have also had an enormous impact on food **packaging**. Now there are a variety of convenient packages which assist the **dispensing** of food and reduce food **wastage**. Tubes, taps (e.g. on juice casks), single-serve portions, blister packs, cook/heat-in packs, easy-opening devices and so on all increase the **appeal** of food and the opportunity to consume it in a variety of **circumstances**.

Company profitability

Long-term **sustainability** in the food industry relies on a food company securing a large portion of the market in the products they produce. In order to be **profitable** over time, the company must:

- Provide products which meet market needs better than their competitors
- Operate efficiently (meet costs + profit)
- Secure supplies of appropriate raw materials that are reasonably priced
- Keep ahead of changing market needs (research)
- Utilise promotion strategies that ensure a profile in the appropriate marketplace
- Think and operate globally.

Types of food product developments

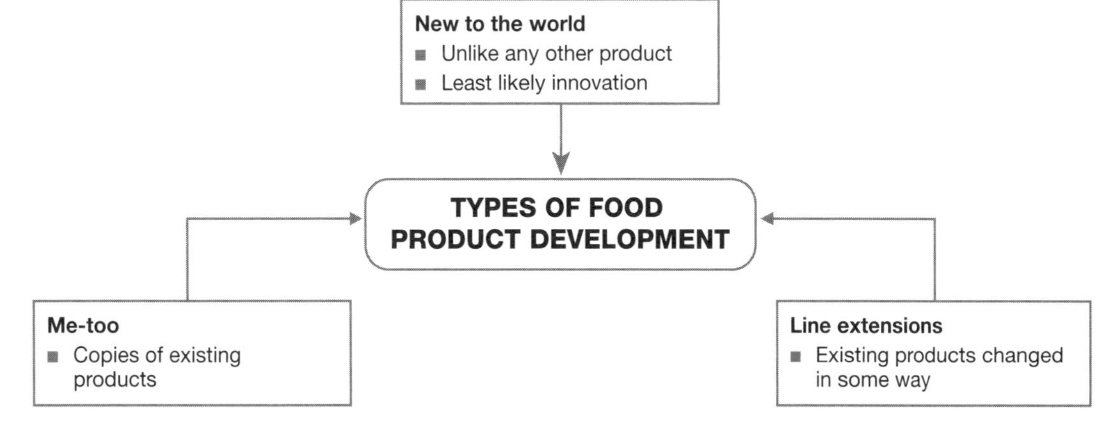

Figure 3.2 Types of food product developments

New to the world products

These are foods which are completely new in concept, unlike any other product in the marketplace. Rarely are food innovations '**new to the world**'. The innovation may be a food, a package or a combination of both. Although few products fit this classification, those that do have the potential to be enormously rewarding for the food manufacturer, such as juice casks, double-up yoghurt and Pringles.

Me-too products

These are **copies** of existing products. A manufacturer may attempt to duplicate the success of another product by developing a similar item. Ingredients, size or shape may be copied within the limits of industrial opportunity. Of the food products developed, most are **me-toos**, and likewise most failed products are also me-toos.

Line extensions

Line extensions are food products which are **changed** in some way to increase their **market share**, such as new flavours, package size or package features. There are many examples of foods developed in this way. At times, the development may be to increase the appeal of the product to health-conscious consumers, such as salt, fat or sugar reduced and complex carbohydrate (fibre) increased.

Activity 7 A PAGE 92

Research the marketplace to identify examples of new to the world products, me-toos and line extensions.

(a) Identify a food which meets a market need.

(b) For that food, identify me-too and line extensions which are available in the marketplace.

3.3 Steps in food product development

Features of food product development in Australia include:

- Food companies are always trying to increase their market share, however, many new products are unsuccessful and only 25% will achieve the results that manufacturers had planned.
- Even though the failure rate is so high, 60% of the products we will purchase in 2015 are not even available yet!
- The development of new food products occurs at many levels; sometimes the steps outlined on the previous page happen concurrently whilst others are isolated steps.
- Food product development is very expensive, so planning and research are important to minimise product failure and maximise profit.
- Line extensions are the most successful new product developments, followed by me-too products and then new to the world products.
- It is important to identify the needs and wants of the target market. The design brief outlines what consumers want and what they are prepared to pay. It also describes the constraints on the producer in producing a new product.

The basic steps in food product development are:

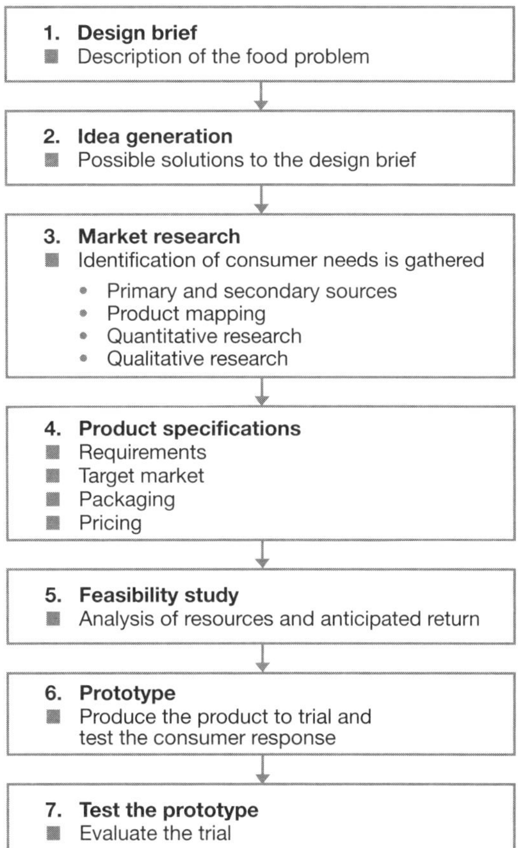

Figure 3.3 Basic steps in food product development

The design brief

This is the **identification** of a food problem and **description** of the situation that requires a solution. A **design brief** may involve a range of food-related problems (not always requiring the development of a new food). Design briefs may relate to:

- Packaging
- Lifestyle
- Health
- Economic situations
- Religious or cultural issues
- Aesthetic appeal.

The design brief should be a **clear statement** that outlines the purpose of the development.

Idea generation

Ideas are then generated to identify possible solutions to the design brief. Initially, the ideas may be extreme, as all possible solutions should be identified. However, **solutions** must fit the company's goals and recognise the company **facilities** (or potential facilities). When the possibilities are identified they should be narrowed down to a smaller number of **feasible** solutions. For every problem many solutions are possible, and the success of the product will depend on the selection of the most appropriate solution.

Market research

Market research needs to be conducted to confirm **consumer needs and wants**. Without consumer support, the solution (product development) will fail.

Research can be:
- **Primary**—gathering data from observations, surveys, sales data and market shares
- **Secondary**—research information and publications (Australian Bureau of Statistics).

The research must be appropriately analysed to ensure that conclusions are valid.

Market research information can be gathered in a variety of ways:

Product mapping

Product mapping identifies **gaps** in the marketplace, demonstrating where there is potential to launch a new product.

Qualitative research

Qualitative research involves checks on **consumer responses** (usually only a **small focus group**) to quality, performance, presentation, pricing and usability. This allows for modification if necessary.

Quantitative research

Quantitative research involves checks on a **large sample** of consumers, often using a questionnaire to seek consumer opinion on five or six items although only one or two will be of interest to the company.

Success relies on clear information about:
- The size of the potential market
- An understanding of market competition
- Available expertise in relation to the product
- Resources/facilities required to develop the product.

Product specifications

If the conditions appear suitable it is then necessary to develop the **product specifications**. This will require:
- A detailed and accurate description of the product and all requirements for production
- Identification of the target market
- Packaging specifications, including sourcing materials
- A pricing strategy.

Feasibility study

This information will be used for the next stage, which is the development of a **feasibility** study with a **product analysis** that includes:
- Anticipated financial return on the projected investment
- A technical assessment of resources available in relation to the anticipated requirements.

Prototype

In order to trial the product it is necessary to develop a production process so that a **prototype** can be developed. A prototype is a **trial product**. Commercial production will not commence until the product has undergone careful **screening**. This may involve:

- Sensory evaluations
- Market tests
- Packaging tests
- Storage tests.

3.4 Marketing plans

Figure 3.4 Marketing plans

Market plans are about increasing sales. A **marketing mix** integrates **four issues** to establish a strategy. The strategy attempts to increase **market share**. Success relies on developing a **market plan** that reflects the needs of the **target market**. Marketing is often described as the four Ps because the key features of an effective marketing strategy all begin with the letter 'P'.

Product

Includes all aspects of the product—ingredients, quantity, flavours, size, packaging and so on. The **product specifications** evolve and are tested during the development of the **prototype**. When the product reaches the production stage, the manufacturer is already very confident as it has been tested on the market and the product specifications match the market demand.

Price

Price can be calculated as **unit price**, or include price variations based on quantity purchased. Prices are often compared to similar products available in the marketplace (**competition**). If pricing is not competitive, the product will fail. Price will fluctuate depending on quality. **Generic** brands have an important position in the marketplace because consumers believe they offer **value for money**. At times consumers are willing to pay a higher price for a product that they believe to be of premium quality. This is often described as a **niche market**.

Place

This is the **availability** of the product in the marketplace, including:

- Where it is sold
- How accessible it is for consumers.

Food items are available from a wide variety of outlets. The important requirement is that the item is readily available to the **target market**.

Promotion

Methods of **promotion** are used to increase consumer awareness of the product and encourage changed consumer behaviour to make initial and then **repeat purchases**. Numerous strategies can be used; for example billboards, television advertising, in-store promotions such as taste testing, free samples and so on. The strategy used must reach the appropriate target market.

The **marketing mix** needs to reflect the stage of the product in its **life cycle**. What is appropriate during the introductory phase may not be as effective during the growth phase. Likewise, strategies used at **maturity** and **decline** will vary accordingly. Unless the marketing mix changes during the product's life cycle, resources will be wasted and mismanagement will result.

Activity 8

Select a food product that you are familiar with and:

(a) Identify the target market.

(b) Explain your understanding of how the marketing mix changed during the product's life cycle.

(c) Imagine you are a food manufacturer. Develop a product that meets a consumer need.

Chapter summary

Factors which impact on food product development can be external or internal.

External factors (the macro environment)

There are a number of influences on food product development that are beyond the control of industry:

- The economic environment
- The political environment
- The ecological environment
- The technological environment.

The Australian food industry is very important to the **Australian economy**.

Internal factors (the micro environment)

There are a number of influences on food product development that are **industry or market based**:

- Personnel expertise
- Financial position
- Production facilities
- Company image.

Reasons for food product development

- Health issues
- The environment
- Convenience and cost
- Societal changes
- Technological developments
- Company profitability.

Types of food product developments

- New to the world products
- Me-too products
- Line extensions.

Steps in food product development

The basic steps in food product development are:

1. Design brief
2. Generation of ideas
3. Market research, including:
   - Product mapping
   - Quantitative research } Research can be primary
   - Qualitative research } or secondary.
   - Product mapping
4. Product specifications
5. Feasibility study
6. Development of a production process and prototype
7. Evaluations and testing.

Marketing plans

Marketing plans aim to increase **market share**. A **marketing mix** integrates four issues to establish a strategy that reflects the needs of the **target market**: The four key features (the four Ps) are:

1. Product
2. Price
3. Place
4. Promotion.

Revision questions

Multiple choice A PAGE 93

1. External factors which influence the development of new food products include
   A employment rates, income, size of the market, working conditions.
   B taxation, employment rates, company image, the environment.
   C skill of the workforce, employment rates, salaries, inflation.
   D the value of the Australian dollar, employment rates, the technological environment, taxation.

2. New food products are developed because
   A there is not enough choice in the marketplace.
   B the Australian food supply does not meet market needs.
   C food companies recognise an opportunity to increase profit.
   D there is too much unemployment.

3. Product specifications are
   A instructions about how to prepare the food.
   B details about the ingredients, size and packaging required to manufacture a food.
   C explained in the manufacturer's production flow chart.
   D the legal requirements for producing the food.

4. Promotion is important in the development of new food products because
   A otherwise people will not know about the product.
   B it helps to increase sales.
   C it is part of any marketing strategy.
   D of all of the above.

5. A useful way of testing a new food product is by sensory analysis because
   A it provides important information about consumers' responses to the product.
   B people can make sense of the idea.
   C there is no other method of deciding if the product will be successful.
   D it is the best way of checking that the machinery can make the product successfully.

6. Internal factors that impact on food product development include
   A environmental legislation.
   B imported foods.
   C production facilities.
   D government policy.

7. Increased consumer demand for convenience foods has occurred because
   A convenience foods are cheaper than meals prepared from scratch.
   B it is healthier to eat processed foods.
   C more women are working and have less time to cook.
   D families are influenced by cooking programs on television.

8. A SWOT analysis is useful when developing new food products because it
   - A prevents competitors from developing me-toos.
   - B results in the producer understanding how to advertise their new food.
   - C controls the political environment so that laws cannot impact on food production.
   - D allows the producer to analyse opportunities in the marketplace.

9. A prototype is a
   - A test sample of the food being developed.
   - B copy of a food that already exists in the marketplace.
   - C discarded idea.
   - D food product that has been successfully advertised.

10. A target market is the
    - A place where food is sold.
    - B consumers who are likely to purchase the food.
    - C price range at which the food is likely to be sold.
    - D group of people who decide on the food company's image.

Short-answer responses A PAGE 93

11. (15 marks)

    You have been asked to develop a marketing strategy for a new frozen fruit drink.

    (a) Describe your target market.

    (b) What type of food product is this frozen fruit drink? Justify your choice.

    (c) Plan a suitable strategy for the marketing of this frozen fruit drink that includes:

    (i) Product planning

    (ii) Price structure

    (iii) Place and distribution system

    (iv) Promotion program.

Structured extended response A PAGE 93

12. (15 marks)

    (a) Explain why most new food products are line extensions.

    (b) Describe how you would develop a new food product to meet a consumer need.

Activity 1

Term	Meaning
Decline	Gradual loss (of market share)
Design brief	Description of a solution to an identified need
Ecological	Biological link between organisms and their environment
Feasibility	Investigating the probability of success
Line extensions	Products that resemble another item except for a change to a characteristic, (e.g. flavour, size)
Macro environment	The broad surroundings
Market research	Gathering of information which provides a description (data) about the marketplace
Market share	Section or percentage of sales that a product commands
Maturity	Maximum growth or development
Me-toos	Copies of other products
Micro environment	Small aspect(s) of the surrounding
New to the world	A product which has never been seen before
Packaging	The container which holds an item(s)
Primary	The original or first-hand
Profitable	Able to make a financial gain
Promotion	Activities which aim to further the growth of an item
Prototype	A sample or model
Qualitative	Relates to qualities or characteristics
Quantitative	Relates to amounts or measurements
Screening	To check the ability or capability
Secondary	Not the original
Sensory evaluation	To make judgements about sound, smell, taste, sight, feel
Specifications	Statement of particular requirements
SWOT	To identify the strengths, weaknesses, opportunities and threats related to a particular investigation
Target market	Anticipated or identified consumers

Activity 2

(a) Purchase cheaper cuts of meat, shop for specials, purchase generic brand products, buy foods in season, eat smaller portions of protein food.

(b) Increased consumption of take-away and partly prepared foods, purchase more convenience foods and less primary foods, eat more tender cuts of meats.

(c) Eat out more often, purchase more convenience foods, consume more luxury food items.

Activity 4

Product line	Health issue	Food and eating strategies
High-fibre	■ Bowel cancer	■ Wholegrain breads and cereals ■ Fresh fruits and vegetables ■ Primary rather than processed foods
Low-fat	■ Obesity ■ Coronary heart disease	■ Reduced-fat dairy foods ■ Lean meats
Low-salt	■ Hypertension (high blood pressure)	■ Primary rather than processed foods ■ Manufactured foods which are salt-free or reduced salt ■ Avoid use of salt in cooking and at the table
Low-sugar	■ Obesity ■ Dental caries	■ Avoid confectionery, cakes, biscuits, desserts, jams and high sugar drinks
Cholesterol-free	■ Coronary heart disease	■ Select plant foods in preference to animal foods (plants do not produce cholesterol)
Nutrient-enriched	■ Vitamin or mineral deficiencies	■ Select manufactured foods which have food labels stating vitamins, minerals, proteins or carbohydrates have been added

Activity 5

Bread is a staple food that has developed in response to a variety of dietary concerns. There are many varieties now available which target a range of health issues as well as personal taste. For example:

- **Whole-wheat** and **whole-grain** breads (rye, oats, barley, corn, rice) are available for consumers wanting the high fibre content of whole-cereal grains. These products are preferred to avoid health problems related to low-fibre diets (constipation, hiatus hernia, varicose veins, haemorrhoids).
- **Low-kilojoule breads** are available for those people trying to maintain a healthy weight range which would avoid a variety of health problems (heart disease, high blood pressure, diabetes).
- **Lite** breads may contain less fat and or sugars. Again they may target overweight people. They may also address the related health concerns of a high fat intake (overweight, high blood cholesterol, stroke, heart disease).
- **Enriched** breads have had fibre or nutrients added. These may include calcium, protein and a range of vitamins and minerals. These breads may be popular because of concerns about a range of health issues (osteoporosis, anaemia, constipation).

Increased concern about health has resulted in more attention being given to food labelling. Bread labels feature information about:

- The number of kilojoules per slice
- Preservatives, artificial colours and flavours
- Quantity of fat (especially trans fats)
- Presence of Omega-3
- Fibre content
- Sugars
- Cholesterol
- Gluten.

Activity 6

- Women working
- More single parent families
- Double income families
- Greater mobility, more time spent away from home (motor vehicle usage)
- Foods are more accessible
- Increased choices (competition)
- Technological advances—less manual labour tasks (quicker, more time-efficient)
- Labour-saving devices
- Busy lives
- Increased value of leisure time

Activity 7

(a) For example, instant single-serve soups.

(b) Me-toos:

- Campbell's Velish tetra pack soup has been developed to provide a product which is intended to duplicate the flavour and quality of a soup made from scratch at home. The packaging is convenient in that it is easily opened and dispensed and can be simply heated to provide a convenient solution for consumers wanting a high-end product.
- Heinz Soup To Go has been developed to provide a convenient food that can be simply heated, served and eaten from the container. The product was developed in recognition of the social change that has seen an increasing number of meals eaten away from home. Convenience, time and cost are important considerations in the development of this food product.

Line extensions:

Continental Cup-a-Soup varieties include:

- Lite products have been manufactured with reduced kilojoules in response to the increased demand for healthy food products.
- Gourmet soups are marketed at a higher price and described as superior in quality and flavour. Products are promoted and sold to consumers demanding superior products.
- Classic soups are the traditional soups that have been available for generations or duplicate the 'home'-style soups of the past (e.g. beef and vegetable, tomato).
- Asian soups are increasingly popular as a result of Asian migration and overseas travel. Consumers have demanded new flavours due to exposure to this style of eating.
- Home-style soups are a convenience food offered as a substitute for homemade soups which are time consuming to prepare.
- Croutons are manufactured and packaged to save the time and energy required to prepare from scratch. They are frequently packaged in multi-packs to provide convenience and freshness.
- Creamy soups meet a market demand for soups that are satisfying and smooth in texture and flavour. They also frequently provide an opportunity to increase calcium intake important for strong bones and teeth.

The development of me-toos and line extensions is recognition of the complexity of the marketplace. Consumer needs are driven by physical, social and economic circumstances; consequently, a wider variety of solutions are available to meet these needs.

Multiple choice

1. **D**
2. **C**
3. **B**
4. **D**
5. **A**
6. **C**
7. **C**
8. **D**
9. **A**
10. **B**

Short-answer responses

11. (a) A target market is the intended consumers, the group likely to purchase the product.
1 mark
Description of target market

(b) Line extension.
1 mark
Identification of the product type

There are many beverages available in the marketplace. This food product has been changed by retailing it in a frozen format to increase market share.
1 mark
Justify your choice

(c) (i) Product – Details of the product size, flavour, packaging materials, ingredients.

(ii) Price – Linked to quantity. Are there several sizes available? How does price vary? Consider competitors and cost comparisons with other similar products.

(iii) Place/distribution – Consider the target market. Supermarkets, stores that sell similar products (e.g. drinks, ice creams); stores that have complimentary products (e.g. fast food).

(iv) Promotion program – How will you increase consumer awareness? For example free samples, television advertising, magazines, in-store promotions.
2 marks each. Links that relate to the frozen beverage for each of the above will increase the mark from a possible 8 to 12 (i.e. the link is worth 1 mark each).

Structured extended response

12. (a) Explains line extensions *1 mark*. Provides explanation of why line extensions are more successful than new to the world or me-too products. *2 marks*

(b) Identifies design brief. *1 mark*

(c) Describes the steps in the development of food products. *7 marks*

(d) Links the food product to an appropriate consumer need. *4 marks*

Contemporary Nutrition Issues

The decisions people make have social, economic, health and environmental consequences. Raising, investigating and debating contemporary nutrition issues enable individuals to make informed decisions and respond appropriately.

Outcomes

On completing this chapter a student should be able to:

1. Evaluate the relationship between food, its production, consumption, promotion and health (H 2.1)
2. Independently investigate contemporary nutrition issues (H 3.2)
3. Develop, realise and evaluate solutions for a range of food situations (H 5.1).

Activity 1 A PAGE 115

Following is a list of terms you should know as a result of completing this module.

Term	Meaning
Allergies	
Cardiovascular	
Food intolerance	
Functional foods	
Heredity	
Malnutrition	
Nutrients	
Nutrition	
Over-nutrition	
Phytochemicals	
Probiotics	
Under-nutrition	

The HSC Examination

In the HSC Examination students could expect to answer questions on contemporary nutrition issues to the value of approximately 25% of the paper. Questions may require students to integrate knowledge, understanding and skills developed through studying the entire course. It is therefore difficult to predict the weighting of marks in each section of the paper.

The paper may include:

- Multiple-choice questions
- Short-answer questions which may include parts
- A structured extended-response question which includes two or three parts with one part worth at least 8 marks

OR

- An extended-response question (approximately 600 words).

4.1 Diet and health in Australia

Physical effects and economic costs of malnutrition and diet-related disorders

Malnutrition is a condition where one or more **nutrients** are not supplied to the body in correct amounts.

Over-nutrition	Under-nutrition
Diet contains an excess of one or more nutrients. Examples include fats, carbohydrates (sugars) and protein.	One or more nutrients are lacking in the diet. Examples include carbohydrates such as fibre and minerals (i.e. iron, calcium) and water.

Diet-related disorders are those conditions not just related to nutrient intake, such as **diabetes and food allergies**.

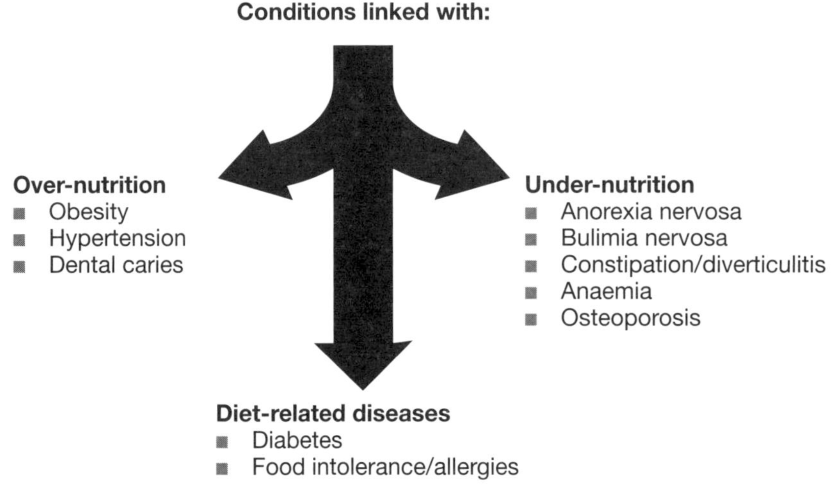

Figure 4.1 Diet-related conditions linked with malnutrition and disease

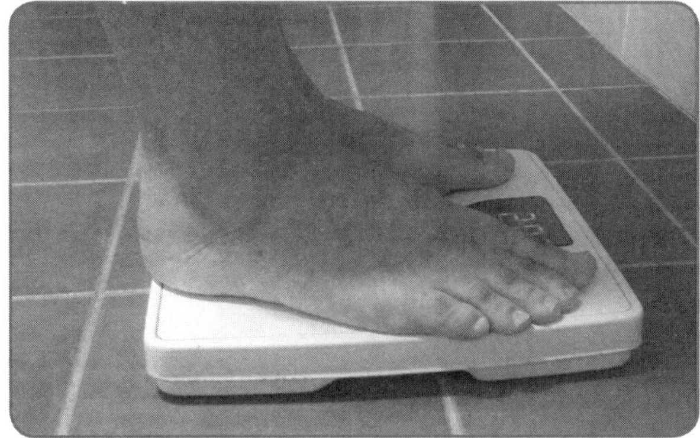

Activity 2 A PAGE 115

For each condition on the following two tables, write a list of five food modifications to include in a diet to prevent the condition.

Condition	Causes	Physical effects	Economic effects
Low-fibre diet (Constipation)	Low-fibre diet causes bulking of faeces in intestine—difficult to remove	■ Hernia ■ Varicose veins ■ Haemorrhoids ■ Diverticulitis—the inflammation of the diverticula* in the intestine causing abdominal pain, nausea and high temperature ■ Colon cancer can result from a low-fibre diet—with less movement of the faeces, wastes remain in the intestine longer and there is increased danger that they come into contact with the cells lining the walls, especially if they contain cancer-causing substances	■ Cost of treatment ■ Employment—sick leave ■ Length of time for treatment ■ Medical facilities
Cardiovascular disease (Coronary heart disease)	Hardening of the arteries (arterio-sclerosis) is caused by a collection of fat (cholesterol) along artery walls, resulting in narrowing of arteries, restricting blood flow to the heart	■ Heart attacks can result from restricted oxygen supply to heart muscle ■ Strokes can also occur if blockage is in the brain	■ Most expensive disease in terms of amount of people suffering that need to be treated ■ Expensive in terms of medication and surgery costs ■ Recuperative time is long ■ Time off work is extensive
Obesity (Excess weight)	When energy intake is greater than energy expenditure	■ Excess energy is stored in adipose tissue (fat) ■ Extra workload for cardiovascular system ■ Fatigue ■ Stress on body joints ■ Reduced body temperature control ■ Associated conditions: ● Diabetes ● Gall bladder disease ● Cardiovascular disease ● Respiratory disorders	■ Cost of hospital treatment ■ Surgery ■ Medicine ■ Toll on ability to work ■ Absenteeism
Hypertension (High blood pressure)	Excess sodium can cause an imbalance of water in body cells, so the body retains water causing the heart to pump blood under pressure	■ High blood pressure ■ Brain haemorrhage ■ Stroke ■ Heart failure ■ Aneurisms (blockage to the blood flow) ■ Kidney disease	■ Cost of hospital treatment ■ Surgery ■ Medicine ■ Toll on ability to work ■ Absenteeism
Dental caries (Tooth decay)	■ Poor dental practices ■ Diet high in sugar ■ Sugar reacts with bacteria in mouth to produce acid	■ Decay on enamel causes nerve exposure in tooth, then toothache ■ Bad breath ■ Loss of teeth	Cost of dental treatment (i.e. dental fillings, crowns and dentures)

** Diverticula are small protrusions from the wall of the intestine that form due to the build-up of pressure in the intestine as the muscles attempt to remove waste.*

Condition	Causes	Physical effects	Economic effects
Anemia (Low iron in blood)	Low iron count in blood can result from: ■ Blood loss ■ Inadequate intake of iron ■ Reduced absorption of iron ■ Repeated pregnancies	■ Chronic lethargy ■ Headaches ■ Dizziness ■ Heart palpitations as the heart pumps faster to deliver required oxygen needs to muscle	■ Less output due to lethargy ■ Need to treat with blood transfusions ■ Iron supplements
Anorexia nervosa (self-induced weight loss)	Psychological obsession with weight loss which can be triggered by a number of factors: ■ Puberty onset ■ Low self-esteem ■ Influence of media on body image perceptions	■ Muscle wasting ■ Abnormally low body weight ■ Absence of consecutive menstrual cycles ■ Slowing of pulse and basal metabolic rate ■ Low blood pressure ■ Anemia ■ Tooth decay due to HCl acid from vomiting ■ Death in extreme cases ■ Burnt oesophagus	■ Due to major occurrence happening from teens to early adulthood, effect on employment ■ Use of expensive resources to treat the condition (i.e. overseas clinics) ■ Psychological counselling ■ Special wards
Bulimia nervosa (Binge eating)	Anxiety	■ Weight loss not prevalent due to huge intake of food ■ Abuse of laxatives can lead to bowel dysfunction. ■ Tooth decay due to HCl acid from vomiting ■ Burnt oesophagus	■ Counsellors and dietitians need to be assigned in hospital ■ Medical cost ■ Family support and community support groups
Osteoporosis (Porous bones)	■ Long-term deficiency of calcium ■ Insufficient intake of calcium ■ Physical inactivity ■ Menopause ■ Inadequate formation of basic bone structure	■ Reduction of bone mass as calcium lost is not replaced ■ Weakened bones ■ Stooped posture	Medical care needed for elderly due to this condition
Diabetes (maturity onset—middle-aged and over)	■ Body produces no insulin or inadequate amounts of insulin ■ Causes include: ● Obesity ● High fat, low-fibre diet	■ Excess blood glucose causes: ● Diabetic coma in extreme cases ● High urination and thirst ■ Energy source comes from adipose tissue as glucose is not available ■ 'Weight loss' muscle wasting can occur ■ Can cause cataracts and glaucoma (blindness) ■ Can cause blood flow problems ■ Renal failure due to excess use of kidneys	■ High cost of medication ■ Constant testing of sugar levels ■ Health problems that result can be expensive to treat

Nutritional considerations for specific groups

Specific community groups have particular **nutritional needs** dependent on:

- Physical state
- Age
- Health status
- Level of exercise
- Nationality.

Within the Australian community, specific groups with **special requirements** include:

- Adolescent girls
- Pregnant and lactating women/teenagers
- Vegetarians
- Athletes
- Elderly
- Post-menopausal women
- Middle-aged men who live alone
- Individuals with specific health conditions, such as diabetes and **cardiovascular** disease
- Aboriginal and Torres Strait Islander peoples (ATSI).

Two of these groups are outlined in detail below.

Group	Strategies to promote optimum health	Dietary needs	Nutritional diseases associated with this group
Aboriginal and Torres Strait Islanders	■ Early medical intervention ■ Maintenance of health and hygiene programs in indigenous areas ■ Nutritional educational programs that encompass culture with western living ■ Promotion of regular exercise ■ Improved supply of food to remote areas ■ Education—teaching literacy skills so that informed choices can be made ■ Education for growing own crops—becoming self-sufficient	■ Less alcohol ■ Better food supply especially in outer areas ■ Less fat ■ More complex carbohydrates ■ Improved cooking techniques using fresh food	■ Heart disease ■ Cancer ■ Digestive disorder ■ Obesity ■ Alcoholism ■ Lactose intolerance ■ Anaemia ■ Diabetes ■ Low life expectancies compared to non-indigenous males/females
Adolescent girls	■ Education of nutritional needs for adolescents ■ Addressing issues of positive self-esteem and body image ■ Teaching food preparation techniques for a healthy diet ■ Education about healthy food choices ■ Effects of alcohol abuse	■ Balanced diet ■ Low fat intake ■ Increased calcium, iron, protein and B-group vitamins	■ Obesity ■ Anaemia ■ Anorexia nervosa and bulimia ■ Alcohol abuse

Activity 3 A PAGE 115

(a) Outline a strategy to promote optimum health in one of the two groups mentioned on the previous page.

(b) List strategies that schools can use to promote healthy food choices in adolescents.

The role of the individual, community groups, the food industry, government orrganisations and private agencies in promoting health

Agencies responsible for health promotion

Figure 4.2 Agencies responsible for health promotion

The responsibility for diet and diet-related health issues in Australia is a complex issue that remains the responsibility of individuals as well as a variety of other agencies.

Group	Why are they responsible?	How each group is addressing the health issue
Individual	■ Makes choices about personal food consumption ■ May also be in a position where their choices impact on others (e.g. parents)	■ Education is needed to understand the implications of poor food choices on health ■ Demand for improved information about food items (e.g. clearer labelling)
Community groups	■ Have the ability to guide food choices and ultimately influence the food industry ■ The media and technology like the internet have facilitated improved access to nutrition information	■ The provision of promotion information (e.g. osteoporosis campaign promoted by the Australian Dieticians' Association in conjunction with the dairy industry to promote the effects of calcium depletion in the diet) ■ Food endorsement programs (e.g. 'Healthy Heart Tick' by the National Heart Foundation, encourages manufacturers to provide products that can be endorsed so that consumers know they are making healthier choices) ■ 'Healthy Canteen' programs in schools have improved food choices for school-age children

Group	Why are they responsible?	How each group is addressing the health issue
Australian food industry	■ The food industry is responsible for the supply of food from the raw material to the final product ■ Consumers rely on this industry for their daily food supply ■ All sectors of the Australian food industry are aware of the impact that available food products have on health	■ Reduced fat, salt and sugar in foods ■ The provision of fibre-rich foods ■ Supply of an increased variety of foods ■ Improvements in production techniques (e.g. organic farming and safety testing of food products) ■ Improved labelling (e.g. the provision of nutrition information) ■ Consideration of ethical marketing practices ■ Increased provision of foods that are appropriate for healthy methods of cooking (e.g. stir-fry and simmer sauces) ■ Promotion of whole foods (e.g. fruit and vegetables)
Government organisations	■ Government policy is based on the reasoning that it is more cost-effective to educate and promote health and positive food choices than to support an increasingly overburdened health system ■ Poor health as a result of food intake is a huge economic and social cost	■ Identification and use of data to target health problems ■ Improved education ■ The provision of tools that inform choices (e.g. Target for Healthy Eating) ■ Formulation of legislation that controls food production (e.g. use of chemicals, genetic engineering and labelling laws that now require nutrition information as well as ingredient specifications) ■ Quarantine laws that protect the food supply and food industry ■ Formation of Food Standards Australia and New Zealand (FSANZ) and the National Health and Medical Research Council to provide guidelines and information on all aspects of food production so that consumers are supplied with the safest food supply possible ■ Department of Health with the support of community dieticians provides advice on healthy food choices
Private agencies	■ Despite the intervention of numerous other groups, the incidence of diet-related health issues continues to impact on the community at large	■ Weight-loss programs to curb the growing incidence of obesity and associated health problems ■ Exercise programs (e.g. gyms and health clubs to support diet in reducing the incidence of overweight, obesity and other associated health problems ■ Organisations associated with the provision of food for the poor and socially disadvantaged ■ Medical personnel (e.g. dieticians and doctors)

Activity 4 A PAGES 115–116

Select ONE agency and identify ways in which it impacts on improved diet and health.

Production/manufacture of nutritionally modified foods to meet consumer demand

Functional foods are foods that have been **changed** to meet consumer demand. They are designed to assist a person's health in either a **curative or preventative** manner.

A **functional food** may be:

- A known food to which a **functional ingredient** from another food is added
- A known food to which a **functional ingredient new to the food** supply is added
- An **entirely new food** that contains one or more functional ingredients.

Functional foods are categorised as:

- Foods that contain **naturally occurring** preventative or curative substances such as antioxidants
- **Modified processed foods** such as fat-reduced, fibre- or vitamin-enriched
- Foods containing non-nutrients such as **phytochemicals**.

In relation to health, **functional foods** are developed to impart improved physical effects beyond those associated with the known **nutrient** in the food. **Recent developments** in **functional foods** include:

- Development of **immune milk products** for prevention of dental caries
- Use of **phytochemicals** as a protective agent against hormone-dependent cancers
- Use of **milk peptides** to lower blood pressure.

Activity 5 A PAGE 116

Complete the following table:

Functional foods	Significant feature	Potential health benefits
Low-fat foods	Low in total fat or saturated fat	
Oatmeal/oat bran/whole oat products	Beta glucan soluble fibre	
Cereal with added folic acid	Folic acid	
Modified margarine	Plant sterols	
Soy	Soy protein	

Fortified foods

Fortified foods are an example of functional foods. Fortified foods have had nutrients added to the food to **increase the levels that were originally found in the food** (prior to manufacture), such as milk fortified with vitamin D and salt fortified with iodine.

The term 'fortified' can be misused by food manufacturers, such as breakfast cereals to which vitamins and minerals are added (as well as high levels of sugar). Fruit juice companies also make claims of improved nutritional value due to fortification but the high levels of sugar added make these claims questionable.

Sometimes fortified foods are confused with **enriched** foods; however, an enriched food only has **nutrients added to replace the original nutrients lost during processing**.

Fermented foods

Fermented foods includes foods that have sugar fermented to alcohol using yeast. These foods are also **functional foods**.

Other processes include the use of bacteria; for example, lactobacillus to make foods such as yoghurt and sauerkraut. Many pickled or soured foods are fermented as part of the pickling or souring process. Some are simply processed with brine, vinegar or another acid like lemon juice.

Role of 'active non-nutrients'

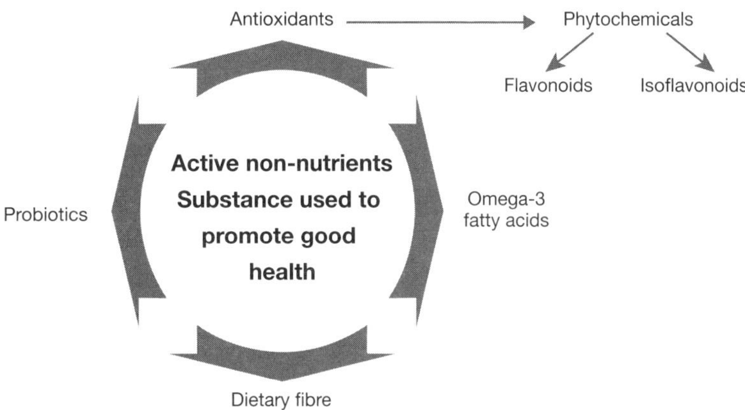

Figure 4.3 The role of active non-nutrients

Antioxidants

An **antioxidant** is a chemical substance that **inhibits oxidation**, and **prevents rancidity of fats and oils**. In the diet, antioxidants can:

- Assist in **cholesterol control**
- Neutralise free radicals, so **reducing risk of cancer**.

Antioxidants are **nutrients** (vitamin E, C and beta carotene) or non-nutrient antioxidants (**flavonoids** and **isoflavonoids**). Sources of antioxidants include fresh fruit, vegetables, legumes and soy beans.

Phytochemicals are non-nutrient antioxidants. Examples are flavonoids and isoflavonoids. **Isoflavonoids** are found in soy beans, seeds, nuts and legumes. **Flavonoids** are colours found in foods—these include anthocyanins that are responsible for most of the red, blue and purple colours in foods, and anthoxans, which are colourless or pale yellow.

Probiotics

Probiotics are micro-organisms of human intestinal origin which are incorporated in foods to improve a healthy gut environment. Their aim is to **reduce gastrointestinal problems** and **improve digestibility** in the gut area. **Probiotics** are also used to:

- Help maintain the immune system
- Contribute to vitamin K foods.

The main micro-organisms added to foods are the lactic bacteria lactobacilli and bifid bacteria. Food products using **probiotics** include yoghurt, sour cream and Yakult.

Omega-3 fatty acids

Omega-3 fatty acid's function is as a precursor of prostaglandins. Prostaglandins are directly involved with the proper functioning of the **cardiovascular** immune system. **Linolenic acid** is a main source of Omega-3 fatty acid. Food sources that provide this include fish oil, soy beans and linseed.

Dietary fibre

Dietary fibre's main functions are to:

- Improve regularity and **prevent constipation** (prevention of constipation can contribute to a reduction in the incidence of polyps in the intestinal tract that can lead to cancer)
- Assist with production of vitamin K and B-group vitamins.

The main food **sources** of dietary fibre include cereals, grains, fresh fruits and vegetables, legumes and fibre-enriched breads and cereals.

Activity 6 A PAGES 116–117

(a) List 10 commercial food products that have been modified in the production process. Look for examples labelled 'fat reduced', 'vitamin enriched', etc.

(b) Explain how modified food products are different to active non-nutrient component foods.

(c) Discuss the role of 'active non-nutrients' in the diet.

The role of supplements in the diet

Diet supplements are **nutrient** substances that are ingested **in addition to** a normal daily diet. The main diet supplements are:

- Vitamin supplements
- Mineral supplements
- Protein supplements.

The main functions of vitamins in the body are:

- Growth and repair of the body
- Vision
- Synthesis of minerals, such as calcium for bones and teeth
- As antioxidants
- Maintenance of cell membranes
- Synthesis of enzymes
- Blood clotting
- Synthesis of protein
- Formation of DNA and RNA in cells
- Energy release from carbohydrates.

The main functions of minerals in the body are:

- Growth and repair of bones and teeth
- Blood clotting
- Nerve and muscle functioning
- Formation of red blood cells
- Enzymic reactions
- Maintenance of blood pressure
- Muscle movement
- Production of energy and amino acids.

The main functions of proteins in the body are:

- Growth and repair of cells
- Formation of enzymes and hormones
- Secondary energy source.

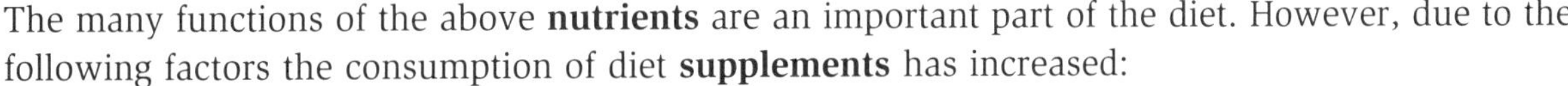

The many functions of the above **nutrients** are an important part of the diet. However, due to the following factors the consumption of diet **supplements** has increased:

- Busy lives often leave **less time for carefully prepared diets and meals**
- Frequent intake of processed food due to the greater disposable income of the average family and **less time for food preparation**
- Increase in food preparation technology has **reduced the use of traditional food preparation skills**.

Instead of making sure they have a balanced diet naturally many people, for the above reasons, choose to supplement their diet with **nutrient pills**. The debate over the value of **supplements** lies in the fact that individuals may be ingesting supplements **without actually needing them**.

The debate over the value of **nutrient** supplements involves these considerations:

- Supplements have been used to **rectify community health problems** (e.g. fluoridation of water to reduce dental caries).
- Excessive doses of supplements can actually lead to **potential health problems**, for example:
  - Excessive doses of fat-soluble vitamins can accumulate in the body and result in **toxic effects**.
  - Excessive doses of water-soluble vitamins are excreted by the body, placing an **overload on the kidneys**.

- Excess protein consumption can contribute to **weight gain** as protein not used is stored in the body as adipose tissue.
- Supplements can be used by people with an **inadequate dietary intake** or those who are physically extended, such as **athletes**.
- Extra additives, such as colours and flavours, are used in supplements to make them **acceptable to the consumer**.
- Better **education** of public **nutrition** could have a flow-on effect, decreasing the need for supplements.
- The added **cost** of supplementing an individual's diet needs to be considered.

Activity 7 A PAGE 117

Debate the role of dietary supplements in a balanced diet.

Activity 8 A PAGES 117–118

List the sectors of the Australian food industry. Identify three ways each sector is supporting improved nutritional food choices.

4.2 Influences on nutritional status

Health and the role of diet in the development of health issues

Individuals are ultimately responsible for their own food choices. However, the link between diet and the following **heredity** conditions does have a significant relationship on the formation of **dietary disorders**.

- **Obesity**
- **Diabetes**
- **Cardiovascular** disease
- Food sensitivity/intolerance/**allergies**.

Obesity

Obesity is the **leading preventable cause of death worldwide** with increasing prevalence in adults and children authorities consider it to be the **most serious health problem of the 21st century**.

Obesity is a medical condition where excess body fat may have an adverse effect on health leading to **reduced life expectancy** and/or increased health problems. It is defined by body mass index (BMI) which is a weight-to-height ratio and is further evaluated by fat distribution via the waist-hip ratio. BMI is related to percentage of body fat and total body fat.

BMI = kilograms/metres^2 (height)

BMI	Classification
<18.5	Underweight
18.5–24.9	Normal weight
25.0–29.9	Overweight
30.0–34.9	Class I obesity
35.0–39.9	Class II obesity
>= 40.0	Class III obesity

Obesity increases the likelihood of diseases such as heart disease, type 2 diabetes, breathing difficulties during sleep (sleep apnoea), certain types of cancers and osteoarthritis.

Obesity is largely caused by **excessive consumption of kilojoules and lack of physical activity**. The most effective **treatment focuses on diet and exercise**.

Measuring obesity in **children** can be problematic due to differences in developmental rates and maturation. It is clear, however, that the incidence of overweight and obesity in children has accelerated—25% of boys and 23.3% of girls were either overweight or obese (*NSW School Physical Activity and Nutrition Survey 2004*).

Adult statistics come from self-reported BMI data from the 2004–2005 National Health Survey (perhaps less reliable as people tend to overestimate their height and underestimate their weight):

- 32.6% of adults were reported as overweight
- 40.5% of males and 24.9% of females were overweight
- 16.4% of adults were obese
- 17.8% of males and 15.1% of females were obese.

As age increases overweight and obesity tends to increase (in the 55 to 64 years age group, 72% of males and 58% of females were overweight or obese).

Diabetes

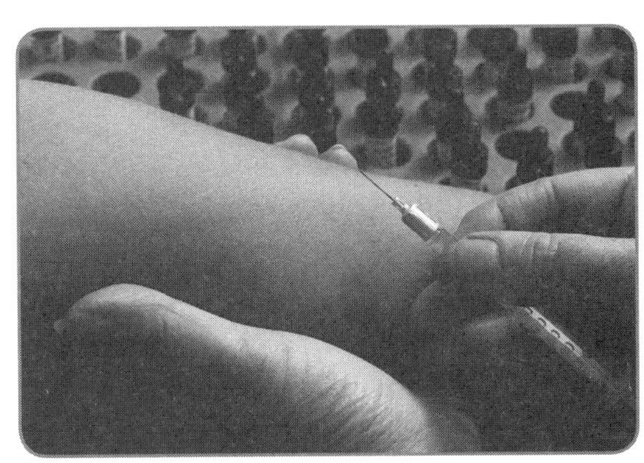

Diabetes is a **disease of the pancreas gland**, where the body is unable to use sugar normally. **Insulin** is a hormone released from the pancreas and is used by the body to regulate sugar levels by moving glucose from the bloodstream to cells. If **not enough insulin is produced**, blood glucose rises, **causing physical side-effects** related to eyes, kidneys, heart, nerves and blood vessels.

Diet can play the following roles in diabetes:

- **Obesity or excess weight** can contribute to diabetes. This condition can cause the pancreas to become inefficient in its production of insulin.
- When first diagnosed, depending on severity, some diabetics can control their condition with the use of a balanced diet. However, many diabetics do become **insulin-injected dependants**.
- For medicated diabetics, awareness of diet is extremely important as they need to **digest enough carbohydrate for the injected insulin to work**. If they don't do this, **hypoglycemia** can occur,

to a point that can cause physical effects such as fainting, trembling and unconsciousness. Such a condition needs to be **treated with a quick injection of carbohydrate** in the form of sugar or a sugar-based food such as jelly beans or fruit juice.

Heredity is also linked to diabetes. If the condition already exists among an individual's family, he or she needs to show awareness by **controlling weight and consuming a balanced diet**.

Cardiovascular disease

Cardiovascular disease (coronary heart disease) occurs due to the **coronary arteries becoming blocked**, stopping blood from being supplied to the **heart muscle**.

The main **nutrient** related to this condition is fat, especially **saturated fats**. Arteriosclerosis is the main cause of **cardiovascular** disease. This is the accumulation of fatty deposits on the interior lining of the artery wall. As the fatty deposits increase, **oxygenated blood flow is reduced**. If the heart does not receive oxygen via blood supply, then **cardiac arrest (heart attack) will result**.

Diet can play the following roles in **cardiovascular** disease:

- Foods high in saturated fats and sugar, if consumed regularly, provide the body with energy. However, any excess of these two **nutrients** is stored as fat and increases blood cholesterol in the body. A **low-fat diet and regular exercise** to use energy would **reduce the risk of cardiovascular disease**.
- **Heredity risks** are evident if there is **family history of cardiovascular disease**. Knowledge of such a history can help raise individual's awareness of their diet, thereby reducing the risk.

Food sensitivity, allergies, intolerances

Food allergy and sensitivity is when the body's immune system **incorrectly identifies** a specific food protein as **foreign**. Once this occurs, **antibodies are produced to fight the foreign body**. The main **nutrient** related to this condition is protein.

The most common foods that produce **allergies** include milk, eggs, gluten, peanuts and fish. Coeliac disease is a reaction to gluten (protein found in wheat). This condition causes tissue changes in the lining of the gastrointestinal tract. This causes a reduction in the **absorption of nutrients**, which can lead to **malnourishment**.

Food intolerance is an individual's reaction to a specific chemical or combination of chemicals in food.

A **food intolerance** is individual, and so can relate to any number of food **nutrients**. An **imbalance** can also be due to **heredity**, which results in the individual having adverse **physical side-effects** from foods.

Food intolerances are mostly related to:

Salicylates	Monosodium glutamate	Lactose
■ Aspirin ■ Benzoates ■ Amines	■ Flavour enhancer	■ Milk sugar
Sources of salicylates: ■ Found in plant foods, fruit, juices, nuts, tea and coffee	**Sources of monosodium glutamate:** ■ Naturally found in tomatoes, mushrooms and cheese	**Sources of lactose:** ■ Found in dairy foods, such as milk, cheese and yoghurt

Strict and correct **dietary management** is the main solution to food allergy and intolerance. The Australian food industry has responded to consumers needs by:

- Reducing the use of **monosodium glutamate**
- Including **additives** on food labelling using the **international coding system**
- Providing food products to suit dietary disorders; for example, **gluten-free products**.

Activity 9 A PAGE 118

(a) Complete the table by describing the relationship between the nutrients carbohydrates, fats and proteins, and their related dietary disorders.

Nutrient	Related dietary disorder	Causes
Carbohydrates	■ Dental caries, food intolerances, obesity ■	■ Over-consumption of simple sugar ■
Fats	■ ■ ■	■ ■ ■
Protein	■	■

(b) Identify examples of food products to assist the management of dietary disorders.

Lifestyle and the effects of cultural and social practices on nutritional status

Nutritional status of the individual can be affected by:

- Lifestyle
- Cultural beliefs
- Social practices.

Lifestyle

Due to the **changing nature of society** many individuals lead a lifestyle that:

- Is **sedentary** due to the introduction of technology for manual tasks and the use of motorised transport systems
- **Lacks time** for consistent exercise patterns
- Is based on leisure choices that **aren't physical**.

People with this type of lifestyle need to have a raised awareness of the necessity to eat a balanced diet and be careful with energy intake. Some **employers** are addressing such issues by providing **exercise equipment** in the workplace, **healthy choice canteens** or participation in **corporate gym membership** as bonus incentives.

Awareness of intake of energy levels could also be raised through the use of **advertising by health groups** and marketers.

Cultural beliefs

Culture in **nutrition** is the acceptance of nutritional practices due to instilled **attitudes, beliefs and values**. Due to the **ethnic diversity of Australian culture**, food habits and choices are a lot more diverse and accepting in nature than in other societies.

Due to the **cooking methods used by other cultures**, Australians are consuming foods that contribute to improved health levels. For example, the use of stir-frying reduces fat intake in cooking procedures or the use of natural herbs and spices lessens the need for artificial flavours in food preparation.

Religion, often through festivals and tradition, influences the nutritional status of some individuals. Depending on the beliefs involved, the **type of food consumed or preparation techniques used** will have a bearing on the diet. Food taboos are those restrictions imposed due to religious beliefs; for example, Seventh Day Adventists are vegetarians. Such food restrictions led to the formation of Sanitarium, a food producer that has improved the **nutritional status** of many individuals due to its production of quality health and vegetarian food products as well as its provision of nutrition and health education.

Social practices

Lifestyle and cultural practices are very much a part of the social practices of **nutrition**. The Australian **environment** dictates to a certain extent the social lives we lead. Many social gatherings, such as barbecues and picnics, are centred on **the use of the outdoors**.

Food producers are catering to such needs by providing food products that can be used in outdoor situations. **Fish and poultry** are being promoted as healthy alternatives to steak for barbecues, and **marinades** that provide variety within this type of meal preparation have been introduced.

The establishment of food courts in **shopping centres** has encouraged consumers to **consume fast food** while participating in a social task such as shopping. The increase in this style of eating has encouraged the use of fast food as a part of a social meeting, which is a concern considering the **quality** of many such foods.

Media and ethical issues related to advertising practices on food consumption

Consumers have become increasingly concerned about nutrition yet are constantly challenged by the confusion of information presented on food labels. In 1995 the National Food Authority released the **Code of Practice on Nutrient Claims in Food Labels and Advertising** (www.foodstandards.gov.au). This was done to prevent consumer deception and promote fair trading. Since this time the wording, illustrations and exaggerated nutritional claims that had been misleading consumers have been reduced. This has also impacted on imported foods as they must also comply with Australian standards in relation to adequate and accurate information.

Unfortunately consumers are often **impulse buyers** selecting food without carefully interpreting the information provided. Therefore despite the strict regulations a lack of time or understanding frequently means that consumers fail to make wise choices.

Advertising as a part of food marketing is used to **persuade consumers to buy a new or existing product**. Advertising organisations conduct research on food consumption patterns to determine what will sell a product. In today's society, a number of factors are considered in food marketing:

- Promotion of health. Due to an increase in dietary disorders and health awareness, through the introduction of diet models such as the Target for Healthy Eating marketers are engaging in **promotion of health** by aiming advertising at **healthy choice and healthy body image**.
- Targeting of the **dual-income family and busy lifestyles** involves the promotion of **convenience foods and fast foods** and represents a major advertising market.
- Targeting the **diverse** Australian culture by the promotion of **ethnic cuisine**.
- Promotion of food products through the involvement of **community support structures** such as reading programs in schools, sports sponsorship and Ronald McDonald House at Westmead Hospital.

The association of fast-food consumption and healthy body image has long been a part of food advertising. Trends show that, due to increased health awareness, the general public does not associate fast-food production with healthy eating. Outlets such as McDonald's are attempting to approach this target by introducing new menus that cater to a busy lifestyle but are still perceived as healthy. McDonald's Shaker Salads are a good example of this.

Ethical issues

At times consumers may well question the ethics of the food industry:

- For example, the increased concern about healthy eating and the apparent unwillingness of the industry to clarify nutritional claims on food labels (see FSANZ www.foodstandards.gov.au)
- The use of prime-time television to promote snack foods to children that contain empty kilojoules (little nutritional value but high in fat, salt and/or sugar)
- Supermarket layouts that ensure 'unhealthy' food options are highly visible (use of end of aisles, children's height and check-out displays)
- Emotive advertising that makes invalid links to healthy lifestyles for unhealthy products
- The use of gimmicks such as free toys and collectables to encourage the purchase of unhealthy food options
- Promotion of 'diet' foods as the answer to weight reduction
- The use of scientific terminology to confuse consumers.

Activity 10

(a) For a food company you have researched, identify five products they have promoted.

(b) What strategies did the company use in their promotion?

(c) How does the company address the issue of health promotion in their advertising?

(d) Imagine you are the advertising executive chosen to sell a low-fat chicken burger for a fast food outlet. Outline the marketing strategies you would use to promote this product in the traditional 'fast food' market.

Activity 11 A PAGE 118

(a) What do the following terms mean on a food label?
  (i) 'Reduced fat'
  (ii) 'Low fat'
  (iii) 'No added sugar'
  (iv) 'Cholesterol free'
  (v) 'High fibre'

(b) Discuss the ethical issues related to the promotion of foods that contain this type of language on the food label.

Activity 12

(a) For a practical you completed in the nutritional unit, discuss the following influences on the choice of that practical:
- Lifestyle
- Culture
- Social factors.

(b) Analyse the nutrient content of the practical you completed.

(c) Adapt the practical to suit a person on a weight-reducing diet.

(d) List five convenience foods that are low in fat and marketed for busy lifestyles.

Chapter summary

The two areas of content that students should know are:

1. **Diet and health in Australia**

- Physical effects and economic costs of malnutrition (under- and over-nutrition) and diet-related disorders.
- Nutritional considerations for specific groups.
- The role of the individual, community groups, the food industry, government organisations and private agencies in promoting health.
- The production/manufacture of nutritionally modified foods to meet consumer demand.
- The role of 'active non-nutrients' in the diet such as photochemicals, probiotics and fibre.
- The role of supplements in the diet.

2. **Influences on nutritional status**

- Health and the role of diet in the development of conditions, including obesity, diabetes, cardiovascular disease, food sensitivity/intolerance/allergies.
- Lifestyle and the effect of cultural and social practices on nutritional status.
- Media and ethical issues related to the effect of advertising practices on food consumption such as the promotion of 'health' foods and 'fast' foods.

Multiple choice A PAGE 119

1. The most serious diet-related health problem in Australia is caused by
   A failing to read nutrition labels on foods carefully.
   B consuming too much food.
   C not eating enough fruits and vegetables.
   D regularly eating fast foods.

2. Fortified foods have been altered to
   A make them appropriate to the treatment of specific illnesses.
   B improve their nutrient content so that they are as healthy as they were before the food was processed.
   C increase the nutrients in the food so that they have more nutrients than were first contained in the food (prior to processing).
   D make the food last longer.

3. Probiotics
   A improve regularity and prevent constipation.
   B replace vitamin c lost during food processing.
   C contain 'good' fats.
   D reduce gastrointestinal problems and improve digestibility.

4. Dietary supplements are important because
   A it is impossible to eat a diet that includes all the nutrients our bodies need.
   B they are an inexpensive way of ensuring good dietary health.
   C they are a safe way of avoiding diet-related diseases.
   D if people have food allergies it may be impossible for them to consume all the nutrients they need.

5. Diabetes is a disease of the
   A pancreas.
   B thyroid.
   C liver.
   D kidneys.

6. Doctors recommend reducing saturated fats in the diet because
   A they contain more kilojoules than other fats.
   B a high consumption of saturated fats can lead to food sensitivities.
   C they lead to atherosclerosis (arteriosclerosis).
   D they are of no nutritional value.

7. As society changes people have become less active because
   A they are more likely to own a car.
   B they work longer hours.
   C they are distracted by leisure activities that are non-active.
   D of all of the above.

8. The most reliable information on good nutrition comes from
   A the media.
   B government education programs.
   C the Australian food industry.
   D weight-control centres.

9. Dietary fibre is contained in
   A cereals and grains.
   B dairy foods.
   C red meats.
   D all food groups.

10. The most likely cause of food allergies are
   A food additives.
   B proteins.
   C carbohydrates.
   D fats.

Short-answer responses **A PAGE** 119

(10 marks)

a) Identify a **diet-related disorder**. (1 mark)

(b) List **two groups** that can be affected by this disorder. (2 marks)

(c) Describe the relationship between **nutrient intake and the identified diet-related disorder**. (3 marks)

(d) Develop strategies to promote prevention of the identified diet-related disorder in one of the identified groups. (2 marks)

(e) Plan a day's menu for one of the specific groups that addresses the specific **dietary** requirements of that group. (2 marks)

Extended response **A PAGE** 120

(15 marks)

Evaluate the **value of dietary supplements** in a balanced diet.

Activity 1

Term	Meaning
Allergies	A physical insensitivity to a substance (e.g. food)
Cardiovascular	Physical system incorporating the heart muscle and circulatory system
Food intolerance	No immune response. Is a reaction in the body after a cumulative build-up of an allergic factor
Functional foods	Foods that provide additional positive health factors in addition to the nutrients present
Heredity	Process by which characteristics are passed on through genes
Malnutrition	A physical condition that is a result of one or more nutrients lacking in the diet
Nutrients	Chemical substances that are food structures
Nutrition	Scientific study of food consumption patterns and nutrient use
Over-nutrition	Excess of one or more nutrients in the diet
Phytochemicals	Antioxidants that come from non-nutrient plant chemicals
Probiotics	Micro-organisms of human intestinal origin used in foods to improve an individual's gut flora
Under-nutrition	Inadequate intake of one or more nutrients in the diet

Activity 2

For example, hypertension:

- Reduce salt intake
- Reduce fat intake
- Increase fibre consumption, such as wholegrain breads and cereals
- Increase consumption of fruits and vegetables
- Reduce alcohol intake.

Activity 3

Target group is adolescent girls. Strategies could include:

- Providing greater variety of healthier food choices in the canteen
- Delivering nutrition information in subjects such as Food Technology and PD, Health and PE
- Having area health personnel speak to school groups
- Asking the student body to survey students on their food choices and make recommendations for improvement
- Providing a breakfast service for students who do not eat breakfast at home.

Activity 4

Government

The federal, state and local governments try to improve the diet and health of Australians in many ways:

- Curriculum in schools that focus on healthy eating and diet-related health issues
- Department of Health programs and information services
- Websites that provide information aimed to educate, such as food standards and food labelling (www.foodstandards.gov.au)

- Research into the relationship between diet and health
- Healthy school canteen policy
- Community dieticians
- Health centres and hospital programs
- Subsidised health care
- Quality control in the food industry, such as quarantine inspections and health surveillance officers.

Activity 5

Functional foods	Significant feature	Potential health benefits
Low-fat foods	Low in total fat or saturated fat	■ Reduced risk of cancer Reduced risk of heart disease
Oatmeal/oat bran/ whole oat products	Beta glucan soluble fibre	■ Reduces cholesterol
Cereal with added folic acid	Folic acid	■ Reduces risk of neural infant defects
Modified margarine	Plant sterols	■ Supports normal healthy cholesterol levels
Soy	Soy protein	■ Reduces risk of heart disease

Activity 6

(a) Bread (e.g. Wonderwhite), milk, cheese, eggs, poultry, beef, yoghurt, margarine, rice, infant formulae.

(b) Modified food products are foods that have substances added to them to improve their nutritional characteristics. Active non-nutrients are those substances that are not essential in the diet but can improve functioning of the body. They can be added to foods or are consumed in their natural state; for example, dietary fibre can be consumed both these ways.

(c) Summary of the following:

Antioxidants
An antioxidant is a chemical substance that inhibits oxidation and prevents rancidity of fats and oils. In the diet, antioxidants can:

- Assist in cholesterol control
- Neutralise free radicals, so reducing risk of cancer.

Antioxidants are nutrients vitamin E, C and beta carotene. Non-nutrient antioxidants are flavonoids and isoflavonoids.

Sources of antioxidants include fresh fruit, vegetables, legumes and soy beans.

Phytochemicals are non-nutrient antioxidants.

Examples are flavonoids and isoflavonoids.

Isoflavonoids are found in soy beans, seeds, nuts and legumes.

Flavonoids are colours found in foods—these include anthocyanins which are responsible for most of the red, blue and purple colours in foods, and anthoxans which are colourless or pale yellow.

Probiotics
Probiotics are micro-organisms from human intestinal origin which are incorporated in foods to improve a healthy gut environment. Their aim is to reduce gastrointestinal problems and improve digestibility in the gut area. They are also used to:

- Help maintain the immune system
- Contribute to vitamin K products.

The main micro-organisms added to foods are the lactic bacteria *lactobacilli* and *bifid* bacteria. Food products using probiotics include yoghurt, sour cream and Yakult.

Omega-3 fatty acids

Omega-3 fatty acids function as a precursor of prostaglandins. Prostaglandins are directly involved with the proper functioning of the cardiovascular immune system.

Linolenic acid is a main source of Omega-3 fatty acid. Food sources that provide this include fish oil, soy beans and linseed.

Dietary fibre

Dietary fibre's main functions are to:

- Improve regularity and prevent constipation
- Contribute to a reduction in the incidence of polyps in the intestinal tract that lead to cancer by preventing constipation
- Assist with production of vitamin K and B-group vitamins.

The main food sources of dietary fibre include cereals, grains, fresh fruits and vegetables, legumes and fibre-enriched breads and cereals.

Activity 7

Advantages:

- Can support people with food intolerances or food allergies
- Assist people who participate in specialised activities that may have specific nutrient needs
- Useful to replace nutrients lost due to cooking, processing or less than fresh foods.

Disadvantages:

- Unnecessary cost when a balanced diet is consumed
- Excess nutrients can be harmful to health (see examples in guide)
- Many water-soluble nutrients in excess to needs will be excreted, putting a strain on the kidneys
- Increase consumption of colours, flavours and fillers
- Can decrease motivation to eat a balanced diet.

Activity 8

Agriculture and fisheries

- Improving crop strains in nutritional standards
- Advertising and production of lean meats, fish, fruits and vegetables
- Increasing variety of fresh fruits and vegetables.

Food service and catering

- Greater variety of restaurants providing healthier alternatives in food choice
- Providing varied serving sizes in meals
- Improving provision of weight loss services by providing catered meals in the home.

Food retail

- Providing larger fruit and vegetable section in supermarkets, so providing greater variety
- Increasing number of health food products
- Increasing variety in bakeries providing appealing breads and cereals which are low-fat, low-salt and high in dietary fibre.

Food processing/manufacturing

- Providing alternative fat-reduced foods
- Improving convenience food by bringing production in line with Healthy Heart Tick guidelines
- Providing functional foods.

Activity 9

(a)

Nutrient	Related dietary disorder	Causes
Carbohydrates	■ Dental caries, obesity ■ Food intolerances ■ Constipation, bowel cancer ■ Diabetes	■ Over-consumption of simple sugar ■ Over-consumption of food (in excess of energy needs) ■ Lack of dietary fibre ■ Inability to metabolise sugar
Fats	■ Obesity ■ Cardiovascular disease ■ Hyperetension	■ Over-consumption of fats ■ Over-consumption of saturated fats ■ Over-consumption of fats
Protein	■ Obesity	■ Over-consumption can contribute to obesity

(b) Increase wholegrain breads and cereals (complex carbohydrates), chicken, fish, fresh fruits and vegetables.

Activity 11

(a) (i) 'Reduced fat': The food must contain 25% less fat than the regular product and at least 3 grams less fat per 100 grams of food (or 1.5 grams of fat per 100 grams if it is liquid).

(ii) 'Low fat' or 'low in fat': The food has no more than 3 grams less fat per 100 grams of food (or 1.5 grams of fat per 100 grams of liquid).

(iii) 'No added sugar': The food has no added sugar (sucrose and other simple sugars), honey, malt extract, glucose syrup or fruit syrup.

(iv) 'Cholesterol free': The food must have less than 3 grams of cholesterol per 100 grams and must meet the requirement for a low-fat claim or have a saturated fat content below a certain level. It is important to remember cholesterol-free does not mean 'fat free'.

(v) 'High fibre': The food must contain more than 3 grams of fibre per 100 grams.

(b) As the 'silent sales person' labels can be unclear and misleading. Food manufacturers have traded on the mystifying claims represented on food labels. A good example of this is 'lite' or 'light' which has very little meaning. It could mean it is light in colour or an ingredient. It does not mean that the food is low in fat. For example, 'lite' potato chips do not contain less fat; they have simply been sliced more thinly. Likewise a label that has 'no added salt' or 'no added sugar' isn't necessarily low in salt/sugar. It is important for the consumer to read the food carefully to determine the level of the ingredients.

Generally a food that is labelled 'toasted' has had fat added. 'Oven baked' also means nothing at all and again the nutrition label is the source of valuable information about fat content. The term '90% fat free' actually means 10% fat. 'All natural' does not necessarily mean 'good food'; fats and oils are 'natural' but high in fat.

The Heart Foundation tick does not necessarily mean the food is 'healthier' that other food options. It simply means that the food with the tick has met Heart Foundation guidelines for total fat, saturated fat, salt, sugar and (where appropriate) fibre. Some companies do not make use of the tick for financial reasons.

Ethically it is inappropriate to trade on people's ignorance or their inability to spend time deciphering nutrient tables on food to find accurate information about food selections. The use of misleading terms deliberately attempt to persuade consumers to make ill-informed purchases is also an ethical issue.

Multiple choice

1. **B**
2. **C**
3. **D**
4. **D**
5. **A**
6. **C**
7. **D**
8. **B**
9. **A**
10. **B**

Short-answer responses

(a) Anaemia.
1 mark
Identifiction of one diet-related disorder

(b) Two groups affected would be **pregnant women and teenage girls**.
2 marks
List two groups (1 mark each)

(c) Anaemia is caused by a low iron count due to blood loss or reduced absorption of iron as in pregnant women.
3 marks
2 marks for discussion of how disorder results
1 mark for linkage to types and amount of nutrient intake

(d) Strategies to prevent anaemia in teenage girls would include:
- Education in schools and media of the need for iron in the teenage years
- Teaching food preparation techniques so that teenage girls are aware of healthy cooking alternatives for iron-rich products such as red meat
- Promoting food products to teenagers that would encourage iron intake, such as red meat and iron-fortified cereals.

2 marks
For two strategies:
1 mark each for describing strategy and for discussion of how it could prevent disorder

(e) Day's menu:
- Breakfast
  - Iron-enriched cereal and milk
  - Orange juice
- Lunch
  - Roast beef and salad sandwich on wholegrain bread
  - Glass milk
  - Yoghurt
- Dinner
  - Stir-fry lamb with Chinese vegetables including Chinese spinach
  - Fruit salad.

2 marks
1 mark for menu plan
1 mark for discussion of how it addresses specific dietary requirements

Extended response

A scaffold of the answer is as follows:

Students need to understand the term 'evaluate', that is, discuss both sides of the statement.

Paragraph one

What is a dietary supplement?

- A **dietary supplement** is a **nutrient** substance that is ingested in addition to a normal daily diet, such as vitamin tablets and protein energy drinks.

What is a balanced diet?

- A balanced diet is composed of a **variety of foods from the core food groups**, that is, breads and cereals, fruits and vegetables, meat and dairy products. The larger part of a balanced diet should be composed of breads and cereals and fruits and vegetables. If a true balanced diet is consumed then the necessity or **redundance of diet supplements** is dependent on external factors.

Paragraph two

Factors that may cause the need for diet supplements:

- **Medical conditions**; for example anaemia raises the need for iron supplements, dental caries cause the need for fluoride in water
- Busy lifestyles leave **less time** for adequate diet planning and food preparation
- An increase in the consumption of processed foods which are often high in fat and low in other essential nutrients due to the **disposable incomes** of dual working couples/families
- **Special needs** such as those of athletes or vegetarians must be met.

Paragraph three

Supplementation is **wasted** or can cause health problems when:

- Excess doses of fat-soluble vitamins (i.e. A, D, E, K) can accumulate and result in **toxic effects**
- Excess doses of water-soluble vitamins are usually excreted so financially they are a waste for the consumer; excess excretion can also place **overload on kidneys**
- Excess protein consumption can contribute to weight gain, as when it is unused the protein is stored as adipose tissue
- Supplementation can also be difficult for consumers financially. There is no case for the extra money spent on supplements when a balanced diet could be bought for less. Technology options such as internet shopping provide busy people with the means to buy wholesome food, even when they lack time
- Better education of basic nutrition principles could have a flow-on effect, so reducing the need for supplements.

Paragraph four

Evaluation paragraph. This should sum up the pros and cons as outlined above.

15 marks

Definition of a balanced diet (2 marks)

Definition of dietary supplements (2 marks)

Linking supplements to diet (1 mark)

Evaluation of dietary supplements (10 marks in total)

- *5 marks for benefits/positive discussion*
- *5 marks for drawbacks/negative discussion*

5 – Sample HSC Examination

Chapter overview

This chapter contains a full sample HSC Examination which can be completed under exam conditions. Full answers are provided, plus useful HSC Exam tips.

HSC Examination tips

Marks

☑ Markers examine what you know. Marks are not taken away—they are only given. So, if unsure, write it and try for a mark. No half-marks are given.

☑ Markers try to help you. They look at everything on the paper, so if you are short of time, jot down:

- a plan
- points
- a mind map
- a diagram

to gain marks.

Time allocation

☑ **Read** the whole paper in the five minutes allowed.

☑ Allocate **time** to gain the best marks. Suggested times to answer each section are printed on the HSC exam paper.

☑ Make your points **clearly** and concisely to avoid repetition and overlong answers. Do not rewrite the question in your answer.

Exam terms

☑ **Link** your answer to the question. Take time to read the question and think about what it is asking. How can your knowledge be linked to the question?

☑ Clearly understand the key terms published by NESA (see page 138).

☑ Look out for all parts to questions. Make sure you answer **all** of every question.

☑ Try to access as many revision questions as possible to increase understanding of terminology.

Spelling and grammar

☑ Under exam conditions, spelling and grammar mistakes are not penalised. However, care needs to be taken with any questions requiring labelling.

Language

☑ Understand the meanings of the words used in this subject and use them.

Sample HSC Examination
Food Technology Operations

General instructions

- Reading time—5 minutes
- Working time—3 hours

Section I **(20 marks)**

- There will be objective response questions to the value of 20 marks.

Section II **(50 marks)**

- There will be approximately 6 short-answer questions.
- Questions may contain parts.
- There will be approximately 14 items in total.
- At least 4 items will be worth 4 to 6 marks.

Section III **(15 marks)**

- There will be one structured extended response question.
- The question will have two or three parts, with one part worth at least 8 marks.
- The question will have an expected length of response of around four pages of an examination writing booklet (approximately 600 words) in total.

Section IV **(15 marks)**

- There will be one extended response question.
- The question will have an expected length of response of around four pages of an examination writing booklet (approximately 600 words) in total.

Source: *Assessment and Reporting in Food Technology Stage 6* © 2009 NSW Education Standards Authority

Notes

- Total marks for this paper is 100 marks.
- A sample multiple choice answer page is included for Section I.
- For practice purposes, use your own A4 paper to answer questions for Sections II, III and IV.
- There is approximately equal weighting of each of the four core strands across the examination as a whole. Questions may require students to integrate knowledge, understanding and skills developed through studying the entire course, rather than focusing on a particular core strand.

  Source: *Assessment and Reporting in Food Technology Stage 6* © 2009 NSW Education Standards Authority
- Answers are given on pages 132–138.

Section I—MULTIPLE CHOICE (20 marks)

Attempt Questions 1–20 in Section I
Use the multiple-choice answer sheet

Multiple-choice questions: answer sheet

Select the alternative A, B, C or D that best answers the question. Fill in the response oval completely.

Sample: 2 + 4 = (A) 2 (B) 6 (C) 8 (D) 9

A ○ B ● C ○ D ○

If you think you have made a mistake, put a cross through the incorrect answer and fill in the new answer.

A ⊗ B ● C ○ D ○

If you have changed your mind and have crossed out what you consider to be the correct answer, then indicate this by writing the word *correct* and drawing an arrow as follows:

A ⊗ B ⊗ C ○ D ○

correct (arrow pointing to B)

1. A ○ B ○ C ○ D ○
2. A ○ B ○ C ○ D ○
3. A ○ B ○ C ○ D ○
4. A ○ B ○ C ○ D ○
5. A ○ B ○ C ○ D ○
6. A ○ B ○ C ○ D ○
7. A ○ B ○ C ○ D ○
8. A ○ B ○ C ○ D ○
9. A ○ B ○ C ○ D ○
10. A ○ B ○ C ○ D ○
11. A ○ B ○ C ○ D ○
12. A ○ B ○ C ○ D ○
13. A ○ B ○ C ○ D ○
14. A ○ B ○ C ○ D ○
15. A ○ B ○ C ○ D ○
16. A ○ B ○ C ○ D ○
17. A ○ B ○ C ○ D ○
18. A ○ B ○ C ○ D ○
19. A ○ B ○ C ○ D ○
20. A ○ B ○ C ○ D ○

Section I—Multiple-choice questions (20 marks)

Answer all questions on the answer sheet provided.
Allow approximately 30 minutes for Section I.

1. Malnutrition is a result of
   A an unbalanced diet.
   B the media influencing people to control their weight.
   C poverty.
   D consuming too many processed foods.

2. Which one of the following is not a sector of the Australian food industry?
   A agriculture and fisheries
   B food service and catering
   C food retail
   D immigration

3. A target market is
   A the place where sales occur.
   B the anticipated consumer group.
   C the group of vendors that control supply.
   D a specific location where sales are checked.

4. A company's market share is determined by
   A productivity.
   B competition.
   C legislation.
   D the *Trade Practices Act*.

5. Sensory evaluation is
   A a prediction of product acceptability.
   B a method of verifying the prototype.
   C a judgement about the flavour, colour and texture of a food.
   D important for quality assurance.

6. The Code of Practice on Nutrient Claims in Food Labels and Advertising
   A gives imported foods an unfair advantage over local foods.
   B makes food manufacturers comply with labelling legislation.
   C ensures that consumers are informed about the nutritional content of food.
   D is an international requirement on all food labels.

7. The global economic crisis has impacted on the Australian food industry because
   A food has become more expensive.
   B the Australian dollar is not competitive in world trade.
   C food manufacturers have no control over external factors.
   D people prefer imported foods.

8. Environmental issues are very important in food production because
   A food manufacturers are required to comply with legislation.
   B the Greens are an important political lobby group.
   C it is an important strategy for cost reduction.
   D in the future Australia aims to produce predominantly organic foods.

9. Manufacturers have introduced material minimisation in packaging to
   A improve package strength.
   B reduce waste and cost.
   C reduce emissions in the atmosphere.
   D change packaging shape.

10. The diverse Australian culture impacts on the food supply because
   A many food retailers are unable to supply imported foods.
   B the Australian eating habits are dictated by the traditional British diet.
   C people are reluctant to try new foods.
   D migrants like to prepare their traditional foods.

11. Market research:
   A checks consumer responses to food products
   B identifies the size of the potential market.
   C identifies the resources and facilities needed to produce a food product.
   D does all of the above.

12. Fortified foods
   A have had nutrients added to the food to increase the nutrient content originally found in the food.
   B are low in fat, salt and sugar.
   C have nutrients added to replace the original nutrients lost during processing.
   D are foods where sugar ferments to alcohol using yeast as the fermenting agent.

13. Genetic engineering is used in food production for which of the following reasons?
   A to reduce the cost of production
   B to improve the nutritional value of food
   C to increase resistance of crops to weeds, pests and diseases
   D all of the above

14. Line extensions are popular developments in the food industry because:
   A they are easy to develop.
   B manufacturers can easily copy other foods on the market.
   C they are always very innovative.
   D they increase the appeal of an already successful product.

15. Increased employment opportunities in the Australian food industry resulted from
    - A reduced use of convenience foods.
    - B limited food choice.
    - C improvements in food manufacturing technologies.
    - D increased tertiary education.

16. The main function of AQIS is to
    - A impose tariffs on imports.
    - B work with Australian Customs and Australia Post to protect Australian agriculture from contamination.
    - C control labelling of agricultural products.
    - D develop food standards.

17. The type of production system used by food manufacturers is mainly dependent on
    - A the nature of the product and scale of production.
    - B the cost of the product.
    - C consumer demand.
    - D the availability of raw materials.

18. Ecologically sustainable food production methods are used to
    - A deplete natural resources.
    - B improve fertility of soil.
    - C decrease the cost of food production.
    - D increase the amount of farms in the agriculture sector.

19. A quality management technique required by law in food manufacture is
    - A HACCP.
    - B quality assurance.
    - C occupational health and safety.
    - D quality control procedures.

20. A function of mineral salts is to improve texture and mouth feel of processed foods. Another additive that would assist this function would be
    - A humectants.
    - B vegetable gums.
    - C bleaching agents.
    - D antioxidants.

Section II—Short-answer questions (50 marks)

Answer ALL questions on the examination paper in the spaces provided.
Allow approximately 90 minutes for Section II.

21. (a) List the reasons for food product developments. (2 marks)

(b) Explain how quality management procedures impact on food production. (5 marks)

22. Using an example, discuss the purpose of conducting a SWOT analysis. (4 marks)

23. (a) Outline a marketing strategy for a specific food product. (3 marks)

(b) Discuss ethical issues that you considered in developing this plan. (4 marks)

24. (a) Identify one organisation in the Australian food industry and name the sector to which the organisation belongs. (2 marks)

Organisation:

Sector:

(b) Explain the following in relation to the organisation you have identified. (4 marks)

(i) Level of operation

(ii) Quality control

(c) Discuss the impact of government policy and legislation on the identified organisation. (6 marks)

25. (a) Select ONE environmental issue and explain how the Australian food industry has responded to the issue. (3 marks)

(b) Evaluate the effect of one emerging technology on the Australian food industry. (6 marks)

26. (a) Identify changes in Australian lifestyle that have impacted on food product developments. (2 marks)

(b) Describe food products that have been developed to meet these changes in lifestyle. (4 marks)

(c) 'Food manufacturers must respond to societal changes if they are to remain profitable.' Evaluate this statement. (5 marks)

Section III

Structured extended-response question (15 marks)

Answer ALL parts of the question in the examination booklets provided.
Allow approximately 30 minutes for Section III.

27. (a) Describe one major Australian dietary disorder. (2 marks)

(b) Explain the relationship between nutrient intake and the disorder that you identified in part (a). (5 marks)

(c) Explore strategies that could be implemented to reduce the impact of this disorder on the health of Australians. (8 marks)

Section IV

Extended-response question (15 marks)

Answer the question in the examination booklets provided.
Allow approximately 30 minutes for Section IV.

28. 'A greater variety of food products, environmental awareness and changing social needs have led to diverse packaging techniques for food.' Discuss this statement.

Answers Sample HSC Examination

Section I

1. **A** Malnutrition is a condition whereby one or more nutrients are not supplied to the body in the correct amounts. This may result in under- or over-nutrition.
2. **D** The four sectors of the Australian food industry are food service and catering, agriculture and fisheries, food retail and food processing.
3. **B** A target market is the consumers likely to consume/use the product.
4. **B** The success of a product is determined by a number of factors including the size of the potential market, available expertise in relation to the product, resources and facilities required to develop the product, and an understanding of market competition.
5. **C** Sensory evaluation is used by marketing departments to confirm acceptance of a product by the use of sensory testing such as taste and sight.
6. **C** This Code is intended to prevent consumer deception and promote fair trading. It aims to control the wording, illustrations and exaggerated nutritional claims that mislead consumers.
7. **C** The economic environment is an external factor that is beyond the control of industry.
8. **A** There is increased concern for the environment, so legislation is in place to ensure that manufacturers operate with due regard for the environmental impact of their production.
9. **B** Material minimisation is the reduction of materials in packaging. This may occur by reducing the use of secondary packaging and making the primary package stronger.
10. **D** Australia is a country of ethnic diversity. People are influenced by their attitudes values and beliefs about food, many of which originate in another culture.
11. **D** Market research is conducted to determine consumer needs and wants. It includes primary and secondary research (both qualitative and quantitative).
12. **A** Fortified foods have had nutrients added to increase the level found in the food prior to manufacture.
13. **D** Genetic engineering is used in food production to improve quantity and quality of food produced.
14. **D** Line extensions are food products that are changed in some way to increase their market share (for example flavour changes, package size or features).
15. **C** Increased employment opportunities in the Australian food industry have resulted from improvements in food manufacturing technologies. This has led to employment opportunities in computers, for example.
16. **B** AQIS uses a number of strategies to protect Australian agriculture from contamination including working with government bodies.
17. **B** The type of production system used by food manufacturers is mainly dependant on the cost of production. Consumer demand helps to determine the type of product produced and the availability of raw materials determines if the product can be produced.
18. **B** Ecologically sustainable food production methods improve soil fertility.
19. **C** Occupational health and safety is used by manufacturers to ensure a safe working environment the use for workers. It is required by law. HACCP is an example of a quality control procedures system and quality assurance is a guarantee of standards.
20. **B** Vegetable gums influence consistency and texture. Humectants are used to absorb moisture, bleaching agents are used to whiten foods and antioxidants are used to prevent rancidity or discoloration by oxidation.

Section II

21. (a) Health issues
The environment
Convenience and food cost
Societal changes
Technological developments
Company profitability

(b) Quality management includes all aspects of quality control during food production.

- Food production methods, product specifications, choice of breeds and varieties, farming techniques, uses of chemicals such as fertilisers and pesticides.
- Production/manufacture, degree to which primary ingredients are refined, addition (or removal) of ingredients, processing techniques and their impact on nutritional value, flavour, texture, shelf life and appearance. Use of critical control checks during production (HACCP) and in some instances application of occupational health and safety procedures.
- Storage and transport at all stages of production control to create conditions that maintain the quality of ingredients and the final product.

22. A SWOT analysis identifies strengths, weaknesses, opportunities and threats in the marketplace and allows food manufacturers to make informed decisions about the potential success of a production idea.

An example of a SWOT analysis might be the development of a new single-serve instant meal. Market research may investigate the characteristics of the prototype including colour, flavour, texture, appearance, shelf life and nutritional value. Consideration would be given to the size of the meal, the type of packaging (and labelling used). The manufacturer would also need to investigate the size of the target market, their wants and needs, and the competitors already in the marketplace. An examination of the resources and facilities available to make the production commercially viable would also be important. Marketing trials may also identify relevant information about a potentially successful product launch and the continued popularity of the product.

23. (a) A marketing strategy will include all four elements of the marketing mix.

Product—may be an energy food for athletes. The ingredients will need to carefully consider energy needs and depletion of nutrients during training and competition. Consideration will also need to be given to quantity/size, packaging, storage and transport of the food to ensure that it meets the needs of the athlete.

Price—will need to be competitive with similar products in the marketplace. The specific nature of the athlete may mean there are opportunities for a premium product in a niche market.

Place—where will this product be available? How will it be sold? It may be sold through retail outlets, training venues such as gyms, coaching staff, supermarkets or franchise opportunities. The profile in the marketplace will have a huge impact on sales.

Promotion—strategies used to promote the food. These may include free samples, the use of a high-profile credible athlete in advertising, research and study data that reflects the product's benefits, and the integrity of the promotion. The mix of strategies will impact on repeat purchases and growth in sales.

It is essential that the strategy reflects the needs of the target market.

(b) One ethical issue is the inappropriate use of scientific language and health claims that do not accurately represent the energy food. Also, high-profile athletes used in advertising who have questionable standards and/or are motivated to promote products for financial gain as opposed to belief in the product is another issue. Food labelling should comply with the Code of Practice on Nutritional Claims in Food Labels and Advertising. The times and places when promotion occurs is also an ethical issue; for example, a high-energy food for athletes should not be marketed as a 'healthy' product for inactive people.

24. (a) Identify an organisation and correctly identify the appropriate sector for that organisation. For example, the organisation could be Sara Lee and the sector could be food processing.

(b) (i) Identify the correct level of operation for the organisation listed in (a) (i.e. household, small business, large company or multinational). An explanation of the level of operation should be given in relation to the identified organisation.

(ii) Quality assurance and quality control processes are explained in relation to the organisations ability to meet standards expected by consumers.

(c) The impact of government policy and legislation at local, state and federal government level and how they affect the operations of the identified organisation should be evaluated. For example, if Sara Lee was the organisation, the response should include evaluation relating to HACCP (federal government), Occupational Health and Safety Act (state government) and codes for inspection of food premises (local government) and how these impact on the operation of the organisation. The impact of HACCP has been improved food production methods and more efficient production, thereby reducing costs for consumers. For the company it does mean stringent guidelines are in place and heavy fines imposed if HACCP is not properly implemented. The Occupational Health and Safety Act has reduced workplace injury, thereby reducing the cost to the consumer as production is not stopped and so costs related to injuries and replacing personnel are not passed on. Codes for inspection of food premises ensure that foods are produced in hygienic conditions, resulting in safer foods for consumers.

25. (a) An explanation of how the Australian food industry has responded to the impact on the environment should include points on waste management, recycling, production techniques and transport. For example:

Waste management/packaging practices

Recycling is the **return** of raw materials into the manufacturing process to make other useful products. It is essential because landfill space in urban areas is dramatically decreasing. Examples of how the Australian food industry manages waste include:

- Material minimisations, which is a strategy introduced by manufacturers to reduce waste and cost by reducing the weight of packaging but still maintaining packaging strength
- Employing waste management companies to manage waste
- Using packaging made from biodegradable material that will decompose through the action of micro-organisms in a reasonable length of time.

Production techniques

- To **reduce energy costs**, manufacturers need to investigate more efficient methods of production and transport.
- **Gases released into the atmosphere** during the production of such packaging products as glass, for example, contain more carbon dioxide than is released during the production of plastic packages, and plastic products are also recyclable.

Solutions to some of the associated problems include:

- **Reusing waste materials**; for example, some food manufacturers reuse water from processing for irrigation of plant gardens
- **Recycling** waste materials such as packaging
- Producing **lighter and more efficiently shaped packages** so that more can be transported at the same time.

Pollution

Pollution comes in many forms: air, land, water and noise. To reduce the impact of pollution, manufacturers have formed waste and pollution **reduction strategies and recycling programs**. Strategies include:

- Reusing and recycling **water**
- Reusing **effluent** and other waste
- Minimising the **use and weight** of packaging
- The Australian packaging industry **eliminating chlorofluorocarbons** (CFCs) and replacing them with hydrocarbons
- Multinational companies such as McDonalds replacing styrofoam packaging with **paper** to reduce the emission of harmful environmental gases, and to reduce landfill because paper is recyclable whereas styrofoam is extremely slow to break down in landfill (not biodegradable)
- Creating by-products from waste; for example, husks and cobs from frozen corn are used for cattle feed.

Transportation

The type of transportation used in distribution (air, water, rail or road) depends on the type of product, distance to be covered and nature of the product. In respect to impact on the environment, food manufacturers are aware of the flow on in costs. While road and air are quick and efficient, rail is certainly more environmentally friendly. Many food companies are establishing warehouses close to rail lines to make use of this mode of transport. In addition, transport companies have fitted filters to exhausts of transport vehicles to reduce gas emissions.

(b) Evaluation of an emerging technology could relate to genetic engineering, organic farming and new packaging techniques. The response must show for and against between the technology and its effect on the Australian food industry. For example, genetic engineering is improving food quality and changing agricultural practices causing the removal of more traditional practices, which could impact on traditional farmers and their share of the agricultural market. There would be an added cost and need for scientific knowledge that the traditional farmers may not be able to afford. Hence genetic engineering may be controlled by larger multinationals or corporations, reducing market share and making it difficult for small-scale farmers to compete. A positive of genetic engineering could be linked to improved quality and quantity of food produced for consumers.

26. (a) Lifestyle changes include:

- Women working and more single-parent families
- Double-income families
- Challenges to traditional roles of men and women
- Increased interest in meal preparation as a hobby
- Greater mobility
- Technological advances
- Increased lifestyle choices
- Busy lives
- Increased value of leisure time
- Increased life expectancy.

(b) Descriptions should include foods such as single serve (e.g. yoghurts, breakfast cereals), heat-and-eat meals (e.g. tetra pack soup, cup-of-soup, frozen meals, *sous vide*—heat-in-the-pouch complete meals) as well as pre-packaged meal solutions (e.g. salads in a bag), instant foods (e.g. mashed potato) and so on. More meals are eaten away from home (breakfast, lunch and dinner) including options such as takeaway, buffet-style eating and restaurant meals. Beverages also range from alcohol to high-sugar non-alcoholic options (e.g. soft drink, flavoured milks). Demand for gourmet ingredients and foods (e.g. sauces, innovative preparation and presentation methods) has increased, as has the ability to source foods from specialty stores (e.g. multicultural foods, niche market produce). There is also demand for a wide variety of meals (meat and three vegetables replaced by culturally diverse food solutions) and an emphasis on quick meal solutions with an interest in more complicated meals as a leisure/entertaining option).

(c) If food manufacturers are to enjoy long-term sustainability they need to have a sufficient share of the market to remain profitable over time. This will mean:

- responding to market needs
- sourcing quality and reasonably priced ingredients
- identifying food solutions in a manner that competes strongly with competitors
- operating in an efficient manner (costs/profit margin)
- conducting market research to ensure that they are aware of marketplace trends
- operating at a global and local level
- investing in technology that allows them to be progressive in the marketplace
- utilising promotion strategies that alert consumers to the manner in which they are meeting their needs.

Section III

27. (a) For example, **obesity** is the leading preventable cause of death in the world. It is the most serious disease of the 21st century. It is a condition where excess body fat has an adverse effect on health, leading to reduced life expectancy. Obesity is defined by body mass index (BMI) which relates to body fat and total body fat. An obese person has a BMI of 30 or greater. This disease can increase the likelihood of diseases such as heart disease, type 2 diabetes, sleep apnoea, certain types of cancers and osteoarthritis.

(b) Many people are becoming obese because they consume more kilojoules than their body requires hence they accumulate fat. The calorie-dense diet is frequently high in fat and carbohydrates (especially sugar).

Frequently a high-fat diet is also low in complex carbohydrates, therefore the obese person may also suffer from constipation, hiatus hernia, varicose veins, bowel cancer. Excess weight can also lead to other health issues such as heart disease because of the additional strain that the weight places on the heart. If the fat consumed is saturated, problems of atherosclerosis or stroke may also occur.

As people become overweight the risk of developing type 2 diabetes is also increased. Many overweight people have a high intake of sugar which can lead to an increased incidence of dental caries. Additionally blood sugar levels may fluctuate, impacting on health.

(c) **Education** is an important tool in improving food choices. School curriculum that focuses on healthy eating programs and inform about the health implications of obesity could influence the incidence of obesity.

A **campaign** like those used against alcohol and tobacco that stress the harmful effects of **excessive food consumption**. This could include data that informs about reduced quality of life and life expectancy.

The government could also consider increased taxes on high-energy low-nutrient food in the same way that alcohol and tobacco are taxed (to discourage use).

Improved labelling ensures that information about nutrition and energy value of foods is visible and clearly stated. Some foods should have warning signs that inform about the lack of nutritional value and high kilojoule content.

Sedentary lifestyles also lead to overweight. Too many people of all ages do not participate in adequate **physical activity**. Strategies need to be implemented to encourage people to get active, such as the implementation of major exercise programs (e.g. early morning get active programs, regional initiatives that encourage people to participate in exercise) and positive recognition for physically active people.

Primary food products need to be promoted to improve their appeal; currently fresh fruits and vegetables lack a positive image because profit margins prevent meaningful promotional programs.

Another strategy could be the introduction of inexpensive **health programs** to support obese people seeking to lose weight.

Section IV

28. Discussion of this question could be structured in the following way:

Food product variety

Packaging is used to:

- Provide convenience (e.g. dual ovenproof materials for use in microwave and oven, *sous vide* for 'heat and serve' meals and MAP packaging for fresh salads.
- Preserve the food product and provide greater variety throughout the year (e.g. active packaging, canning, MAP packaging and use of foil laminates).
- Communicate, with manufacturers using the package as a marketing tool.
- Protect products and allow for the development of packaging to suit new foods.

Environmental awareness

- Manufacturers are addressing the 'cradle to grave' concept so that overall cost effectiveness of the package is considered before its use.
- Increased use of materials that cause less environmental impact during production of the package (e.g. water-based inks on labels).
- Packages are designed so they can be re-used.
- Lighter packaging materials are developed to make energy use during transport more cost-effective.
- Recyclable materials (e.g. glass) are used, reducing landfill and uses fewer raw materials.
- Lightweight plastic is used instead of glass. This material requires less use of energy in its production and is also recyclable.

Changing social needs

- There is less time for food shopping and preparation, increasing the need for convenience in meal preparation and clean-up (e.g. microwave meals and food products packaged to last longer using active and MAP packaging techniques).

- Dual-income families have increased; they can afford convenience foods which are often more expensive.
- Increased demand for single-serve packages has led to the need to reduce waste of excess packaging by making materials recyclable or reusable.
- Increased market competition between food producers has led to packages being designed as an effective marketing tool improving labelling, packaging size and secondary use of the package so consumers think they are getting value for money.
- Increased health awareness has increased the need for provision of fresh produce to be available for a balanced diet. Hence developments in MAP packaging have resulted to provide longer-lasting primary produce.

Key examination question terms

Term	Definition
Define	Make a statement which states the essential nature or scope of the term
Identify	Recognise and name selected features and examples of a concept, such as the **advantages** of **credit**
Explain	Give reasons or account for by means of detailed explanation—link cause and effect, such as 'the relationship between'
Describe	Show understanding of a concept by use of descriptive words and phrases telling the characteristics or appearance of something
Interaction	A human process in communicating among individuals or groups necessary to form a relationship
Discuss	Construct an argument or case to support or deny a claim
Critically discuss	Discuss showing both sides of the argument and make a judgement
Relationship	Links or connections between two or more phenomena
Explore	In-depth description which does not seek to suggest one solution
Factors	Influences or facts which cause some effect or result
Issues	Ideas or situations about which there is some discussion, contention or disagreement in society
Implications	Possible or sugested outcomes of decisions/actions
Options	Things that are or may be chosen as possible alternatives
Compare	Recognise similarities and differences
Strategies	Actions consciously chosen as part of a plan
Evaluate	Make a judgement after consideration of positive and negative effects/truths of a statement
Justify	Provide sound evidence on which the response is based
Assess	Make a judgement of the qualities/abilities of something to serve a purpose or to influence
Analyse	Examine in detail the structure of a statement/situation, separating it into component parts
Critically analyse	Analyse showing positive and negative instances

Notes

Notes